T0295479

APPLIED AQUACULTURE BIOFLOC TECHNOLOGY

The intent of this book is to provide a detailed and specific set of guidelines for both aquapreneurs and researchers related to the application of Biofloc Technology in aquaculture. The goal of the book is to remove the false claims and untrue stories on biofloc from scientific point of view as it is applied to aquaculture and to provide a guide for its accurate application. This book discusses key issues related to both adoption and practices for aquaculture businesses, how to monitor and assess quality and quantity of biofloc, and how to manage the microbial composition and sludge reduction risk in the fish and shrimp culture. The book works through the specific application of disease and feed management tools for aquaculture from the perspective of this technology. Particular attention is paid on comparing the prototypes of floc development and evaluation on its efficacy in aquaculture. A section on application of biofloc technology in research should assist aquaculture researchers to avoid common mistakes. Additional chapter on design and operation of exsitu bioreactor models provide a sense of more applicability in replacing fish meal.

Prof. S. Felix has 38 years of teaching and research experience in aquaculture. He is the Past President of World Aquaculture Society-Asian Pacific Chapter and he has also served as the Director of the World Aquaculture Society (APC). Prof. S. Felix has served as Second Vice Chancellor of Tamil Nadu Dr. J. Jayalalithaa Fisheries University and also served as Dean of the Fisheries College in Tamilnadu, India.

Dr. M. Menaga is currently working as an Assistant Professor in the Department of Aquaculture at Tamil Nadu Dr.J Jayalalithaa Fisheries University and she's also the Student Director of World Aquaculture Society- Asia Pacific Chapter. She has a M.F.Sc in Aquaculture from Central Institute of Fisheries Education, Mumbai and a Ph.D. in Tamil Nadu Dr.J Jayalalithaa Fisheries University (TNJFU).

APPLIED AQUACULTURE BIOFLOC TECHNOLOGY

S. Felix
Former Vice Chancellor
Tamilnadu Dr. J.J. Fisheries University
Vettar River View Campus,
Nagapattinam, Tamil Nadu 611002
and
M. Menaga
Assistant Professor
Tamilnadu Dr. J.J. Fisheries University
Vettar River View Campus,
Nagapattinam, Tamil Nadu 611002

NARENDRA PUBLISHING HOUSE
DELHI (INDIA)

First published 2022
by CRC Press
2 Park Square, Milton Park, Abingdon, Oxon, OX14 4RN

and by CRC Press
6000 Broken Sound Parkway NW, Suite 300, Boca Raton, FL 33487-2742

CRC Press is an imprint of Informa UK Limited

British Library Cataloguing-in-Publication Data
A catalogue record for this book is available from the British Library

Library of Congress Cataloging-in-Publication Data
A catalog record has been requested

ISBN: 978-1-032-15124-3 (hbk)
ISBN: 978-1-003-24261-1 (ebk)

DOI: 10.1201/9781003242611

Contents

Preface

The expansion of the aquaculture production is restricted due to the pressure it causes on the environment by the discharge of waste products in the water bodies and by its dependence on fishoil and fishmeal. Aquaculture using biofloc technology (BFT) offers a solution to both problems. It combines the removal of nutrients from the water with the production of microbial biomass, which can be adopted *in-situ* and *ex-situ* wise by the culture species as additional food source.We are very much concerned about our contribution in promoting this technology as the science behind the technology has not reached the aquapreneuers adequately and considering the current trend of exploitation of resources and methods of production this technology will be the most ideal culture technique for aquaculture. Biofloc technology in the past decade has taken great leaps forward in efficiency and hence economic viability through a wide range of technological advances has been achieved in several countries. Presently, our ability to understand the environmental costs of industrial farming increases, we are more capable of developing technologies to ensure that farming is more productive and less damaging to the environment. This eco-friendly technology allows us to be more productive, and although we have no certainty that we can and will effectively solve climate change, we still have hope that there will be a future where people will be healthy and fed with nutritious food. We make an unique effort to explore the applications of biofloc technology as we, realise that we are but small fry in a world of much bigger fish, but we are more than hopeful, indeed confident, that biofloc technology has a role to play in the world's future food production. This volume is testament to the knowledge and enthusiasm we have in promoting this green aqua farming technique with a motive of reducing carbon emission in aquaculture. We made strenuous effort and we are sure this would definitely pave a way to learn the applications of the biofloc technology along with the knowhows on operation and maintenance. This Volume-I will agument our capability to steer the microbial aggregation to obtain optimal morphological characteristics (floc size and floc size distribution) to serve as food for the culture species. Our research focusing on the nutrient removal from the water and on the compositional aspects (protein, polyunsaturated fatty acids, lipids, poly-β-hydroxybutyrate, etc) of the bioflocs, was also dealt in this book for the added value for aquaculture.

S. Felix
M. Menaga

CHAPTER 1

APPLIED AQUACULTURE BIOFLOC TECHNOLOGY

Aquaculture as a food-producing sector is recognized for its ability to contribute to the global fish supply. Aquaculture production is projected to rise to 82 million tonnes in 2050 (FAO, 2016). To prevent and treat diseases in aquatic animals, antibiotics have been used for improving aquaculture production. Frequent usage of antibiotics may result in the development of antibiotic resistant bacteria, antibiotic residues in the flesh and the microbial population destruction in the aquatic environment (Marques *et al.*, 2005). This leads to the adoption of various alternative strategies for the antibiotic including the use of pre and probiotics. Sustainable intensification by the adoption of advanced culture systems and technologies becomes inevitable to improve the production and productivity of the sector. One of such advanced technologies is biofloc technology (BFT) also called as aerobic microbial floc (AMF), a minimal or zero water exchange technology, which allows the animals to stock at higher densities. Bioflocs are conglomerates of algae, bacteria, protozoans, faecal matter and uneaten feed which are held together in a loose matrix by the secretions of filamentous microorganisms or by electrostatic attraction (Hargreaves, 2013). Their growth is influenced by various physico chemical factors such as temperature, pH, salinity and aeration of the culture systems. The growth and stability of the flocs formed can also be determined by using different carbon sources and at various C/N ratios. This technology by maintaining the carbon and nitrogen in the culture water uses the dense microbial biomass to strip the ammonia and serves as a nutritional supplement (Schneider *et al.*, 2005). Unlike Recirculatory Aquaculture Systems (RAS) this technology does not require any external filtration rather dense microbial biomass, strips the ammonia in turn to serve as a nutritional source. As biofloc technology deals with bacteria and bacterial products bioflocs might also contain immunostimulatory compounds such as PHB exhibiting possible probiotic effect. The heterotrophic bacteria in biofloc (Halet *et al.*, 2007) produce some natural substances (Dinh *et al.*, 2010; Iyapparaj *et al.*, 2013) and suppress the growth of other pathogenic species like *Vibrio harveyi* (Defoirdt *et al.*, 2007).The various carbon sources such as rice bran, rice flour, and wheat flour have been previously used for the floc development. These carbon sources tend to slowly dissolve in water thereby increasing the suspended solids concentration. The distillery spent wash considered as a major waste from sugarcane industry with a higher dissolving capacity than the complex carbon sources can be used. They have the ability to rapidly release the carbon sources paving the way for the heterotrophic bacteria to assimilate the ammonia with decreased concentrations of TSS thereby maintaining an ideal water quality (Felix *et al.*, 2015). The use of distillery spent wash as a carbon source has been recommended for shrimp as well as GIFT Tilapia (Menaga *et al.*, 2017). The nutrient profile of biofloc ranges from 25 to 50 percent of protein and 0.5 to 15 percent of fat on dryweight basis. Bioflocs are also valuable source of limiting amino acids such as methionine and lysine, vitamins (Vitamin C in range of 0 to 54 μg /g dry matter) and limiting mineral such as phosphorus (Crab, 2010). In aqua feeds, dried biofloc can be used possibly to replace fishmeal or soybean meal as cheaper sources of protein. Extensive and traditional systems with no or little use of fishmeal supplies nutrient-rich materials to the culture water enhancing the growth of algae and other indigenous organisms on which the fish can feed (Naylor *et al.*, 2000).The effluent waters from aquaculture systems were used for *ex-situ* biofloc production in suspended growth bioreactors. The bioflocs produced can be dried and used as a feed supplement for shrimp or fish (Kuhn and Lawrence, 2012).The use of biofloc as a feed has been studied in aquaculture and the uptake of biofloc as feed depends on the nature of species and its feeding ability, size of the animal and floc and the density of floc (Avnimelech, 2009).The findings from the previous study proved that freshwater prawn, shrimp and tilapia uptake the biofloc as the additional protein source indicating that it can be applicable to both freshwater and seawater culture (Crab, 2010; Crab *et al.*,2009, 2010a). Biofloc helps in the potential feed gain with decreased production cost (Craig and Helfrich, 2002) which can be estimated to be in the order of 10– 20% (De Schryver *et al.*, 2008).

DIFFERENT TYPES OF BIOFLOC SYSTEMS

Various biofloc systems have been evaluated in research and are also used in commercial aquaculture. The type includes the system that is sunlight dependant and sunlight independant. The outdoor system includes lined ponds or tanks and indoor system includes lined raceways in green houses for the culture of shrimp or tilapia. In "green water" biofloc systems a conglomerate of algae and bacteria controls water quality. Most of the biofloc in outdoor commercial system use green water. However, some biofloc systems (raceways and tanks) with no exposure to natural light have been installed in closed buildings. These systems are categorized as "brown- water" biofloc systems which involves bacterial processes for controlling water quality (Hargreaves, 2013). Hargreaves (2013) stated the types of phytoplankton addition in the biofloc culture system. They can be added through the water at the time of system start-up or inoculated from the phytoplankton stock. He also stated that in green-water biofloc system, phytoplankton are the major groups involved in maintaining the water quality by assimilation of excess ammonia-nitrogen. Phytoplankton due to is autotrophic nature by involving in photosynthesis enriches the culture system with the oxygen produced. Choo *et al.*, (2015) observed that in indoor tanks ammonia concentrations are mainly controlled by the heterotrophic bacterial population in the biofloc. In indoor system, C/N ratio is increased through supplementation of an organic carbon or reduced protein levels in feed, the heterotrophic bacteria create a demand for nitrogen in the form of ammonia in the water and as organic carbon and inorganic nitrogen are generally taken up by the bacteria controlling the ammonia concentration.

McIntosh (2001) reported the biofloc types based on the colour as green, brown and black and when ingested the animals showed an improved results in both shrimp as well as fishes.

In-situ and *Ex-situ* Culture in Biofloc Technology

In biofloc systems, intake of microbial flocs in *insitu* manner and the supplementation of processed floc as feed ingredient has been evaluated (Kuhn *et al.,* 2009, 2010; Anand *et al.,* 2014). Ju *et al.,* (2008) showed that the free amino acid concentrations like alanine, glutamate, arginine and glycine,which serve as an attractants in shrimp diet (Nunes and Martin, 2004), are known to be present in bioflocs. Bioflocs are considered as food particles by some micro and macro aquaculture organisms since it is suggested that the nutrient levels in bioflocs were similar to the fish commercial diet. Furthermore, in larviculture, application of biofloc technology may provide food source that are easily accessible for the larvae beside the intake of commercial diets, thus serves as an alternative strategy for reducing cannibalism during feeding (Ekasari *et al.,* 2015). In culture ponds or tanks *insitu* bioflocs are formed by management of carbon and nitrogen ratio (C:N) to values of 8:1 or higher by the supplementation of carbon source such as glycerine, sucrose, molasses and calcium acetate. In other way, low protein feeds can also be used to maintain a higher C: N ratio in the culture water. Heterotropic bacteria are the primary components of bioflocs under the conditions of higher C:N ratios for the intake of these bioflocs by shrimp and fish(Kuhn *et al.*, 2010).' Avnimelech *et al.*, (1986) stated that biofloc technology retains waste and converts into food. Avnimelech (1999) found 20% reduction in feed requirement, while culturing tilapia in pond in which animal can utilize the biofloc and digest it as *insitu* food source which was available for 24 hours in the culture system. Hari *et al.*, (2006) showed the stimulation of heterotrophic bacteria for the synthesis of microbial protein by increasing the C/N ratio through external carbohydrate addition to the culture systems. The biofloc technology (BFT) is an advanced technique deployed in aquaculture for maintaining the optimal water quality parameters with the development of heterotrophic microbial biofloc by supplementing carbohydrate source to the culture water (Avnimelech *et al.*, 1989; Avnimelech, 1999; Crab *et al.*, 2007; Crab *et al.*, 2009; Crab *et al.*, 2010a).Sinha *et al.*, (2008) further reviewed the prospective use of biofloc as a functional ingredient in aquaculture and suggested that *insitu* biofloc culture can be an alternative strategy for disease management to other existing applications such as use of probiotic, antibiotic and prebiotic. Hargreaves (2013) reported that *insitu* shrimp culture water, boost the growth of animal and floc since it possess growth stimulating components such as microbial and animal proteins, and can be used as a food source for the grazers like shrimp or tilapia. The formation of *ex-situ* biofloc is carried out in suspended-growth biological reactors where settled solids and nitrate were removed from aquaculture used culture waters. To promote biological activity carbon supplementation can be adopted in the bioreactors. The used culture water from aquaculture ponds can be diverted for biofloc production using biological reactors or fermentors (Kuhn and Lawrence, 2012). Kuhn *et al.*, (2009) prepared *ex-situ* biofloc feed by using sequential batch reactors, these biofloc meal were used as ingredient for juvenile shrimp diet. On the other hand, biofloc meal was produced with activated sludge system by Neto *et al.*, (2015) fed to shrimp at 30% replacement of fishmeal in diet. Hende *et al.*, (2014) also reported the production of biofloc by using pikeperch culture water treated in a pilot reactor. The incorporation of biofloc meal at a rate of 80g/kg to the shrimp diet showed no deleterious effect on growth and survival. Biofloc meal produced from culture waters of super intensive shrimp farm was reported by Bauer *et al.*, (2012). Izquierdo *et al.*, (2006) have demonstrated that the supplementation of the floc to shrimp in mesocosm culture reduced the use of fish oil in pelleted diets. Ju *et al.*, (2008) collected biofloc from marine shrimp culture tanks as whole floc and with several extraction modes as a diet or diet component for shrimp. McIntosh *et al.*, (2000) analysed suspended flocs from commercial shrimp ponds in BFT systems at Belize. They found higher protein levels in suspended floc derived from the floc compared to the commercial feeds. To replace fishmeal or soybean meal in aqua feed, dried biofloc had been proposed as an ingredient as biofloc meals are rich in nutrients while they are also free from mycotoxins levels, anti-nutritional factors and other components that restrict its use in aqua feeds (Emerenciano *et al.*, 2013). Large scale biofloc meal production for use in aquaculture at commercial scale results in the environmental advantages to various ecosystems, thus demand of fish meal/oils an ingredient will be minimized (Kuhn *et al.*, 2010; Bauer *et al.*, 2012).

BIOFLOC AS FEED

Owing to the increased demand of seafood for human consumption, the demand for fishmeal and fish oil used for aqua feed production by feed manufacturers also increases (Péron *et al.*, 2010). Besides its application in aquaculture, fish meal and fish oil are used as feed ingredient for poultry, pigs and other animal husbandry sectors. Aquaculture production relies on capture fisheries, as fish oil and fish meal are important components of the diet of many aquaculture species, both herbivorous and carnivorous fishes. In aquaculture about 5–6 million tonnes of low-value/trash fish were used as feed or feed ingredient (FAO, 2009). FAO (2009) reported that between 1992 and 2006, in aqua feeds the total amount of fishmeal and fish oil used was grown more than three fold from 0.96-3.06 million tonnes and from 0.23 - 0.78 million tonnes, respectively. Based on the feed conversion ratio of major aquaculture species, an average of 1.9 kg of wild fish is needed for 1kg of fish production (Naylor *et al.*, 2000). An increased use of fish rates, protein in terms of fish meal upto 2-5 times has been used in many intensive and semi-intensive aquaculture systems (Naylor *et al.*, 2000). Therefore, recent research has focused on the alternatives of fish oil and fishmeal, using cheaper sources of plant and animal proteins. The use of fishmeal in extensive or any conventional systems was limited due to the growth of algae and other native microorganisms in the culture water compared to intensive and semi intensive systems (Naylor *et al.*, 2000). Thus biofloc technology, was developed to overcome the use of fishmeal as an alternative in intensive and semi-intensive systems. Nitrogenous waste generated by the cultured organisms in biofloc technology is converted into microbial protein through the external supplementation of carbon source and hence *insitu* feed production can be stimulated (Schneider *et al.*, 2005).

BIOFLOC ON BIOSECURITY

In aquaculture development, green water technology has been adopted with varied perceptions. So far, aquaculture has been misunderstood due to the poor handling of antibiotics; which encouraged scientists to discover new strategies for controlling pathogenic infections (Naylor *et al.*, 2000). According to Defoirdt *et al.*, (2011), in treating bacterial disease most antibiotics were no longer effective because of the development of antibiotic resistance. Crab *et al.*, (2010) reported that in aquaculture facilities biofloc technology could be an alternative strategy to defend against bacterial pathogens. In bioflocs some groups of bacteria tend to involve in cell-to-cell communication (quorum sensing) with small signal molecules (Defoirdt *et al.*, 2008). In fact, Defoirdt *et al.*, (2004) reported the expression of virulence factors is disrupted by bacterial cell-to-cell communication to maintain the healthy environment. Studies of Lezama-Cervantes and Michel (2010) showed the increased primary production and beneficial bacterial group promotion in *insitu* biofloc based shrimp culture and to some extent it was proved in Tilapia culture as well. Recently to provide in depth knowledge on disease resistance towards pathogenic infections scientists have hypothesized the presence of immunostimulatory compounds in biofloc (Crab *et al.*, 2012). According to Wang *et al.*, (2008), immuno-stimulants are group of active compounds which includes complex carbohydrates, animal extracts, bacteria and bacterial products, nutritional factors, cytokines, lectins and plant extracts.

BIOFLOC ON PROBIOTIC EFFECT

In contrast to conventional approaches such as antibiotic, antifungal, probiotic and prebiotic application biofloc can be a novel strategy for aquatic health management. In BFT the "possible probiotic" effect against *Vibrio sp.* and ectoparasites either internally or externally was observed. Mostly bacteria and large groups of microorganisms promote this effect. Internally, bacteria and the bacterial products have a similar effect like organic acids acting as an effective biocontrol agent to maintain microbial balance in the gut (Sinha *et al.*, 2008). The frequent addition of carbon source in the water is known to promote polyhydroxyalkanoates (PHA) accumulating bacteria and similar groups of bacteria that accumulate PHA granules. Poly-B-hydroxybutyrate (PHB), a biodegradable polymer, the microbial storage product is the single compound of a whole family belonging to polyhydroxyalkanoates. PHB is produced from soluble organic carbon by various microorganisms such as *Bacillus* sp., *Alcaligenes* sp., *Pseudomonas* sp. and have a role in energy storage and carbon metabolism (Sinha *et al.*, 2008). Meanwhile the presence of PHB in biofloc has been reported upto 16% on dry weight basis (De Schryver *et al.*, 2012). Different carbon sources or carbon substrate structures will result in different types of PHA (De Schryver *et al.*, 2012). Under limiting nitrogen conditions with excess carbon such granules are synthesized (Sinha *et al.*, 2008). The degradation of these polymers by chemical and enzymatic hydrolysis in the animal gut could exert antibacterial activity similar to organic acids and short chain fatty acids SCFAs) (Yu *et al.*, 2005). Enzyme hydrolysis takes place by extracellular depolymerases activity of bacteria and fungi, acting as antimicrobial agents against *Vibrio* sp. infections and stimulates growth and survival of fish larvae and shrimp (De Schryver *et al.*, 2012). The working mechanism of antibacterial activity of PHAs is not well understood (Sinha *et al.*, 2008). As they act similar to SCFA, the speculated working mechanism includes (i) reduction of pH to increase the antibacterial activity (Ricke, 2003); (ii) the pathogenic bacterial growth is inhibited by its interaction on cell membrane structure and permeability, depletion of ATP cellular energy, instability of internal proton balance (Russel, 1992) and (iii) reduced virulence factor expression influencing the gut health (Teitelbaum and Walker, 2002). To maximize PHA content in bioflocs further research is needed for analysing and characterizing their bio-control efficiency in host microbe systems (Sinha *et al.*, 2008). Externally against pathogens the working efficiency of biofloc microorganisms seems to be affected through competition for space, substrate and nutrients. Essentials nutrients such as nitrogen are needed by both nitrifying bacteria vs *Vibrio* sp. restricting their growth inhibiting compounds secreted by beneficial microorganisms. The type of carbon source and light intensity could also reduce the growth of pathogen. Unfortunately in this field only limited information is available. Studies with fish fingerlings (Ekasari *et al.*, 2015) recorded that tilapia (initial weight 0.98 ± 0.1g) reared under BFT exhibited lower degree of parasitical infections in gills as compared to conventional water-exchange system (CW) after 60 days.

BIOFLOC ON WATER QUALITY

In aquaculture to maintain optimum water quality biofloc technology has become a sustainable technique and this can be achieved with the presence of dense heterotrophic bacterial groups with the supplementation of carbon source to the water (Avnimelech *et al.*, 1989; Avnimelech, 1999; Crab *et al.*, 2007; Crab *et al.*, 2009). However, the evidences from the previous studies showed that the type of carbon source influences the culture water quality and nutritional composition of floc (Crab, 2010). According to Avnimelech and Ritvo (2003), only 25% of the feed nutrients are fed by the culture animals leading to higher nitrogen accumulation, particularly total ammonia nitrogen. In aquaculture systems nitrogen pathways that naturally remove ammonia nitrogen includes autotrophic mode of conversion from ammonia-nitrogen to nitrate-nitrogen, Photoautotrophic removal by algae and heterotrophic bacterial removal of ammonia-nitrogen directly to microbial protein (Ebeling *et al.*, 2006). Therefore, the presence of dense heterotrophic microbial load in ponds can speed up the biological removal of organic and inorganic wastes in ponds (Avnimelech, 1986; Azim *et al.*, 2003a).

Invitro Probiotic Properties

Multiple benefits of biofloc in sustainable disease management practices for protecting the culture animal with proper biosecurity measures was compared with the traditional aquaculture practices such as use of probiotics, prebiotics and antibiotics by Sinha *et al.*, (2008).The heterotrophic bacterial dominance established by maintaining C/N ratio in the water with the interventions in using different external carbohydrate sources or by using low protein feeds helps the bacteria to assimilate the ammonia for converting to microbial biomass (Avnimelech, 1999).The heterotrophic bacteria in biofloc (Halet *et al.*, 2007) produce some natural substances (Dinh *et al.*, 2010; Iyapparaj *et al.*, 2013) and suppress the growth of other pathogenic species like *Vibrio harveyi* (Defoirdt *et al.*, 2007).

Polyhydroxy Butyrate (PHB) Producing Bacteria in Biofloc Systems

Biofloc is the conglomeration of bacteria, plankton, algae, faecal wastes and excess feed. PHB producing bacteria are found in bioflocs and levels of PHBs in biofloc ranges from 0.5 -18% of the dry matter (Crab 2010; De Schryver and Verstraete, 2009). The presence of PHB producing bacteria in different ecologies ranging from freshwater to marine water as well as in the estuarine sediments and air was reported by Jendrossek & Handrick (2002). Supplementation of PHB in feed improved growth and survival of *Macrobrachium rosenbergii* (freshwater prawn) larvae (Nhan *et al.*, 2010), *Oncorhynchus mykiss* (rainbow trout) (De Wet 2005), *Dicentrarchus labrax* (European seabass) juveniles (De Schryver *et al.*,2010) and also showed protection against vibriosis in *Artemia franciscana* (Defoirdt *et al.*, 2007), *Eriocheir sinensis* (Chinese mitten crab) larvae (Sui *et al.*, 2012).

Extracellular Polysaccharides (EPS) Producing Bacteria in Biofloc Systems

As sludge from aquaculture ponds is negatively charged they

can interact with EPS, a synthetic polymers in combination with cations to neutralize the surface charges thereby aiding the flocculation and settling (Higgins and Novak, 1997b). However these synthetic polymers when added to the sludge water not only enhance the flocculation and sludge dewatering but also when discharged into the environment affects the soil microorganism as a major disadvantage.

CONCLUSION

To support aquacultural growth, taking into account sustainability, researchers stand for challenges regarding water quality control tools and fish feed technology. This book aimed at evaluating the possibility of using biofloc technology as a sustainable tool to support the necessary future increase in aquaculture production by accomplinning water quality control and immune enhancement through *insitu* and *exsitu* feeding of biofloc. Due to the novelty of the technique, many research fields remain unexplored and further development of those domains is needed. Since so many study directions show a great potential, the objective here was to look at biofloc technology from various research angles to enlighten the aspects of the advantages of biofloc technology. This book was chose to enlighten an in depth research on biofloc technology with respect to the microbial composition and its role in biofloc to improve the over all performance of the aquatic animals.

REFERENCES

Anand, P.S., Kohli, M.P.S., Sundaray, J.K., Roy, S.D., Venkateshwarlu, G., Sinha, A.,Pailan, G.H., 2014. Effect of dietary supplementation of biofloc on growth per-formance and digestive enzyme activities in Penaeus monodon. Aquaculture. 418, 108-115.

Avnimelech, Y., 1999. Carbon/nitrogen ratio as a control element in aquaculture systems. Aquaculture. 176, 227–235.

Avnimelech, Y., Mokady, S., Schroeder, G.L. 1989. Circulated ponds as efficient bioreactors for single cell protein production. Israeli Journal of Aquaculture. 41(2), 58-66.

Avnimelech, Y., 1999. Carbon/nitrogen ratio as a control element in aquaculture systems. Aquaculture. 176, 227–235.

Avnimelech, Y., 2009. Biofloc technology: a practical guide book. World Aquaculture Society.

Avnimelech, Y., Weber, B., Hepher, B., Milstein, A., Zorn, M., 1986. Studies in circulated fish ponds: organic matter recycling and nitrogen transformation. Aquaculture Research.17(4), 231-242.

Avnimelech, Y., Ritvo, G., 2003. Shrimp and fish pond soils: processes and management. Aquacult. 220(1-4), 549-567.

Azim, M.E., Milstein, A., Wahab, M.A., Verdegam, M.C.J., 2003. Periphyton–water quality relationships in fertilized fish ponds with artificial substrates. Aquacult. 228, 169-187.

Bauer, W., Prentice-Hernandez, C., Tesser, M. B., Wasielesky Jr, W., Poersch, L. H., 2012. Substitution of fishmeal with microbial floc meal and soy protein concentrate in diets for the pacific white shrimp Litopenaeus vannamei. Aquaculture.342, 112-116.

Craig, S., Helfrich, L.A., 2002. Understanding Fish Nutrition, Feeds and Feeding (Publication 420–256). Virginia Cooperative Extension, Yorktown (Virginia).4 pp.

Crab, R., 2010. Bioflocs technology: an integrated system for the removal of nutrients and simultaneous production of feed in aquaculture. PhD thesis submitted to Ghent University. 178 pp

Crab, R., Kochva, M., Verstraete, W., Avnimelech, Y., 2009.Bio-flocs technology application in over-wintering of tilapia.Aquaculture Engineering. 40, 105–112.

Crab, R., Chielens, B., Wille, M., Bossier, P., Verstraete, W., 2010a. The effect of different carbon sources on the nutritional value of bioflocs, a feed for Macrobrachium rosenbergii postlarvae. Aquaculture Research. 41, 559–567.

Choo, H.X., Caipang, C.M.A., 2015. Biofloc technology (BFT) and its application towards improved production in freshwater tilapia culture. Aquac. Aquar. Conserv. Legis Int. J of Bioflux (AACL Bioflux). 8(3).

Crab, R., Avnimelech, Y., Defoirdt, T., Bossier, P., Verstraete, W., 2007. Nitrogen removal techniques in Aquaculture for a sustainable production. Aquacult. 270, 1–14.

Crab, R., Defoirdt, T., Bossier, P., Verstraete, W., 2012. Biofloc technology in aquaculture: beneficial effects and future challenges. Aquaculture. 356, 351-356.

Dinh, T.N., Wille, M., De Schryver, P., Defoirdt, T., Bossier, P., Sorgeloos, P., 2010. The effect of poly-β-hydroxybutyrate on larviculture of the giant freshwater prawn (Macrobrachium rosenbergii). Aquaculture.302, 76–81.

Defoirdt, T., Halet, D., Vervaeren, H., Boon, N., Van de Wiele, T., Sorgeloos, P., Bossier, P., Verstraete, W., 2007. The bacterial storage compound poly-β-hydroxybutyrate protects Artemia franciscana from pathogenic Vibrio campbelli. Environmental Microbiology. 9 (2), 445–452.

De Schryver, P., Crab, R., Defoirdt, T., Boon, N., Verstraete, W., 2008. The basics of bioflocs technology: the added value for aquaculture. Aquaculture. 277, 125–137.

Defoirdt, T., Sorgeloos, P., Bossier, P., 2011. Alternatives to antibiotics for the control of bacterial disease in aquaculture. Current opinion in microbiology. 14(3), 251-258.

Defoirdt, T., Boon, N., Sorgeloos, P., Verstraete, W., Bossier, P., 2008. Quorum sensing and quorum quenching in Vibrio harveyi: lessons learned from in vivo work. The ISME journal.2(1),19.

Defoirdt, T., Boon, N., Bossier, P., Verstraete, W., 2004. Disruption of bacterial quorum sensing: an unexplored strategy to fight infections in aquaculture. Aquaculture.240(1-4), 69-88.

De Schryver, P., Boon, N., Verstraete, W., Bossier, P., 2012. The biology and biotechnology behind bioflocs. In Biofloc technology: a practical guide book. World Aquaculture Society (WAS). pp. 199-215.

Dinh, T.N., Wille, M., De Schryver, P., Defoirdt, T., Bossier, P., Sorgeloos, P., 2010.The effect of poly-â-hydroxybutyrate on larviculture of the giant freshwater prawn (Macrobrachium rosenbergii). Aquaculture.302, 76–81.

De Schryver, P., Verstraete, W., 2009. Nitrogen removal from aquaculture pond water by heterotrophic nitrogen assimilation in lab-scale sequencing batch reactors. Bioresource Technology. 100(3), 1162-1167.

De Wet, L., 2005. Can organic acids effectively replace antibiotic growth promotants in diets for rainbow trout Oncorhynchus mykiss raised under sub-optimal water temperatures. Abstract CD-Rom. World Aquaculture Society.9–13.

De Schryver, P., Sinha, A.K., Baruah, K., Verstraete, W., Boon, N., De Boeck, G., Bossier, P., 2010. Poly-beta-hydroxybutyrate (PHB) increases growth performance and intestinal bacterial range-weighted richness in juvenile European sea bass, Dicentrarchus labrax. Appl. Microbiol. Biotechnol. 86, 1535–1541

Emerenciano, M., Cuzon, G., Paredes, A., Gaxiola, G., 2013. Evaluation of biofloc technology in pink shrimp Farfantepenaeus duorarum culture: growth performance, water quality, microorganisms profile and proximate analysis of biofloc. *Aquaculture international.*21(6), 1381-1394.

Ebeling, J.M., Timmons, M.B., Bisogni, J.J., 2006. Engineering analysis of the stoichiometry of photoautotrophic, autotrophic, and heterotrophic removal of ammonia- nitrogen in aquaculture systems. Aquacult. 257, 346–358.

Ekasari, J., Rivandi, D.R., Firdausi, A.P., Surawidjaja, E.H., Zairin Jr, M., Bossier, P., De Schryver, P., 2015. Biofloc technology positively affects Nile tilapia (Oreochromis niloticus) larvae performance. Aquaculture. 441, 72-77.

FAO, 2016. The State of World Fisheries and Aquaculture(SOFIA): Contributing to food security and nutrition for all. Food and Agriculture Organization (FAO), Rome. 200p.

Felix, S., Antony Jesu Prabhu, P., Cheryl Antony, Gopalakannan, A., Rajaram, R., 2015.Studies on nursery rearing of Genetically Improved Farmed Tilapia (GIFT) in Biosecured raceway systems in India. J. Indian Fish. Assoc. 42, 1-10.

FAO (Food and Agriculture Organization of the United Nations). 2009. Fishmeal market report — May 2009. Food and Agriculture Organization of the United Nations Globefish.

Hargreaves, J.A., 2013. Biofloc production systems for aquaculture.1-11.

Halet, D., Defoirdt, T., Van Damme, P., Vervaeren, H., Forrez, I., Van de Wiele, T., Boon, N.,Sorgeloos, P., Bossier, P., Verstraete, W., 2007. Poly-β-hydroxybutyrate-accumulating bacteria protect gnobiotic Artemia franciscana from pathogenic Vibrio campbelli. FEMS Microbiology Ecology. 60, 363–369

Hari, B., Kurup, B.M., Varghese, J.T., Schrama, J.W., Verdegem, M.C.J., 2006. The effect of carbohydrate addition on water quality and the nitrogen budget in extensive shrimp culture systems. Aquacult. 252, 248-263.

Hende, S.F.D., Claessens, L., Muylder, E.D., Boon, N., Vervaeren, H., 2014. Microbial bacterial flocs originating from aquaculture wastewater treatment as diet ingredient for Litopenaeus vannamei (Boone). Aquac Res. 47, 1075 – 1089

Halet, D., Defoirdt, T., Van Damme, P., Vervaeren, H., Forrez, I., Van de Wiele, T., Boon, N.,Sorgeloos, P., Bossier, P., Verstraete, W., 2007. Poly-β-hydroxybutyrate-accumulating bacteria protect gnobiotic Artemia franciscana from pathogenic Vibrio campbelli. FEMS Microbiology Ecology. 60, 363–369.

Higgins, M.J., Novak, J.T., 1997b. Characterization of exocellular protein and its role in bioflocculation, J. Environ. Eng. 123, 479–485.

Iyapparaj, P., Maruthiah, T., Ramasubburayan, R., Prakash, S., Kumar, C., Immanuel, G., Palavesam, A., 2013.Optimization of bacteriocin production by Lactobacillus sp. MSU3IR against shrimp bacterial pathogens. Aquatic Biosystems.9 (1), 12

Izquierdo, M., Forster, I., Divakaran, S., Conquest, L., Decamp, O., Tacon, A., 2006. Effect of green and clear water and lipid source on survival, growth and biochemical composition of

Pacific white shrimp Litopenaeus vannamei. Aquacult. Nutr.12, 192- 202.

Ju, Z.Y., Forster, I., Conquest, L., Dominy, W., 2008. Enhanced growth effects on shrimp (Litopenaeus vannamei) from inclusion of whole shrimp floc fractions to a formulated diet. Aquac.Nutr.14, 533e543.

Jendrossek, D., Handrick, R., 2002. Microbial degradation of polyhydroxyalkanoates. Annual Review of Microbiology. 56(1), 403-432.

Kuhn, D.D., Boardman, G.D., Lawrence, A.L., Marsh, L., Flick, G.J., 2009. Microbial floc meal as a replacement ingredient for fish meal and soybean protein in shrimp feed. Aquaculture. 296, 51-57.

Kuhn, D.D., Lawrence, A.L., Boardman, G.D., Patnaik, S., Marsh, L., Flick, G.J., 2010. Evaluation of two types of bioflocs derived from biological treatment of fish effluent as feed ingredients for Pacific white shrimp, Litopenaeus vannamei. Aquaculture. 303, 28-33.

Kuhn, D.D., Lawrence, A., 2012. Ex-situ biofloc technology. In: Biofloc Technology a practical guide book. (2nd edition).edited by Avnimelech, Y., 217-230. Journal of the World Aquaculture Society, Baton Rouge, Louisiana, USA.

Kuhn, D.D., Lawrence, A., 2012. Ex-situ biofloc technology. In: Biofloc Technology a practical guide book. (2nd edition).edited by Avnimelech, Y., 217-230. Journal of the World Aquaculture Society, Baton Rouge, Louisiana, USA.

Kuhn, D.D., Lawrence, A.L., Boardman, G.D., Patnaik, S., Marsh, L., lick, G.J., 2010. Evaluation of two types of bioflocs derived from biological treatment of fish effluent as feed ingredients for Pacific white shrimp, Litopenaeus vannamei. Aquaculture. 303, 28-33.

Kuhn, D.D., Boardman, G.D., Lawrence, A.L., Marsh, L., Flick, G.J., 2009. Microbial floc meal as a replacement ingredient for fish meal and soybean protein in shrimp feed. Aquaculture. 296, 51-57.

Lezama-Cervantes, C., Paniagua-Michel, J., 2010.Effects of constructed microbial mats on water quality and performance of Litopenaeus vannamei post-larvae. Aquacultural engineering. 42(2), 75-81.

Marques, A., Dinh, T., Ioakeimidis, C., Huys, G., Swings, J., Verstraete, W., Dhont, J., Sorgeloos, P., Bossier, P., 2005.Effects of bacteria on Artemia franciscana cultured in different gnotobiotic environments. Applied and Environmental Microbiology.71, 4307-4317.

McIntosh, D., Samocha, T.M., Jones, E.R., Lawrence, A.L., McKee, D.A., Horowitz, S., Horowitz, A., 2001.The effect of a bacterial supplement on the high-density culturing of Litopenaeus vannamei with low-protein diet in outdoor tank system and no water exchange. Aquac Eng. 21, 215–227.

Menaga, M., Felix, S., Gopalakannan, A., 2017.Identification of Bacterial Isolates from Aerobic Microbial Floc Systems by 16sr DNA Amplification and Sequence Analyses. Indian Vet. J. 94, 24 – 26.

McIntosh, D., Samocha, T.M., Jones, E.R., Lawrence, A.L., McKee, D.A., Horowitz, S., Horowitz, A., 2000.The effect of a bacterial supplement on the high-density culturing of Litopenaeus vannamei with low-protein diet in outdoor tank system and no water exchange. Aquac Eng. 21, 215–227

Neto, H.S., Santaella, S.T., Nunes, A.J.P., 2015. Bioavailability of crude protein and lipid from biofloc meals produced in an activated sludge system for white shrimp, Litopenaeus vannamei. Rev. Bras. Zootecn. 44, 269- 275.

Naylor, R.L., Goldburg, R.J., Primavera, J.H., Kautsky, N., Beveridge, M.C.M., Clay, J., Folke, C., Lubchenco, J., Mooney, H., Troell, M., 2000.Effect of aquaculture on world fish supplies.Nature. 405, 1017–1024.

Nunes, A.J.P., Martins, P.C.C., Gesteira, T.C.V., 2004. Carcinicultura ameaçada: produtores sofrem com as mortalidades decorrentes do vírus da mionecrose infecciosa (IMNV). Pan.Aquic. 14, 37–51.

Naylor, R.L., Goldburg, R.J., Primavera, J.H., Kautsky, N., Beveridge, M.C.M., Clay, J., Folke, C., Lubchenco, J., Mooney, H., Troell, M., 2000.Effect of aquaculture on world fish supplies.Nature. 405, 1017–1024.

Nhan,D.T., Wille,M., De Schryver,P., Defoirdt,T., Bossier,P., Sorgeloos,P., 2010. The effect of poly –â -hydroxybutyrate on larviculture of the giant freshwater prawn Macrobrachium rosenbergii. Aquaculture. 302, 76-81.

Péron, G., Mittaine, J.F., Le Gallic, B., 2010. Where do fishmeal and fish oil products come from? An analysis of the conversion ratios in the global fishmeal industry. Marine Policy. 34, 815–820.

Ricke, S.C., 2003. Perspectives on the use of organic acids and short chain fatty acids as antimicrobials. Poultry science.82(4), 632-639.

Russel, J.B., 1992. Another explanation for the toxicity of fermentation acids atlow pH: anion accumulation versus uncoupling. Journal of Applied Bacteriology. 73, 363–370.

Schneider, O., Sereti, V., Eding, E.H., Verreth, J.A.J., 2005. Analysis of nutrient flows in integrated intensive aquaculture systems. Aquaculture Engineering. 32, 379–401.

Sinha, A.K., Baruah, K., Bossier, P., 2008. Horizon scanning: the potential use of biofloc as an anti-infective strategy in aquaculture -an overview. Aquaculture Health International.13, 8-10.

Sui, L., Liu, Y., Sun, H., Wille, M., Bossier, P., De Schryver, P., 2012.The effect of poly-β-hydroxybutyrate on the performance of Chinese mitten crab (Eriocheirsinensis Milne-Edwards) zoea larvae. Aquaculture Research. 1-8

Teitelbaum, J.E., Walker, W.A., 2002. Nutritional impact of pre- and probiotics as protective gastrointestinal organisms. Annual review of nutrition. 22(1), 107-138.

Wang, Y.C., Chang, P.S., Chen, H.Y., 2008. Differential time-series expression of immune-related genes of Pacific white shrimp Litopenaeus vannamei in response to dietary inclusion of â-1,3-glucan. Fish Shellfish Immunol. 24(1), 113–121

Yu, K.S.H., Wong, A.H.Y., Yau, K.W.Y., Wong, Y.S., Tam, N.F.Y., 2005. Natural attenuation, biostimulation and bioaugmentation on biodegradation of polycyclic aromatic hydrocarbons (PAHs) in mangrove sediments. Marine pollution bulletin. 51(8-12), 1071-1077.

CHAPTER 2

IN-SITU AND *EX-SITU* BIOFLOC ON IMMUNE RESPONSE OF GIFT TILAPIA

The present study is aimed to investigate the effect of biofloc intake on Genetically Improved Farmed Tilapia (GIFT), developed within the system and its influence as feed supplementation on water quality, growth performance, immunological parameters, antioxidant status, immune gene expression, and its resistance to *Aeromonas hydrophila* infection. GIFT Tilapia juveniles of 5.1 g (±0.05) were stocked at a density of $15/m^3$ in lined ponds of 300 m^2 in triplicates for 180 days. The experimental groups consisted of T1- biofloc developed within the culture systems (*insitu*), T2- biofloc supplementation in fish feed (*exsitu*) and C- Control without biofloc. Distillery Spent wash was used as a carbon source to maintain the C/N ratio of 10:1 for floc development in T1. Free CO_2, pH, BOD, dissolved oxygen, alkalinity, Calcium and Magnesium ions, Nitrate, Nitrite and Ammonia were found to be significantly different between the treatments and control throughout the experiment. The immunological (Serum protein, Respiratory burst test (RBT) and Myeloperoxidase) and antioxidant indicators (Glucose, Superoxide dismutase (SOD) and catalase) were found to be significantly higher in T1 at the end of the trial. Increased weight gain, specific growth rate, survival and decreased feed conversion ratio was found in T1 when compared with the other experimental groups. Real time quantitative PCR analysis revealed that there was no folded expression of the immunological genes such as Metallothionein gene, Cathepsin L, Toll like receptor 7, Interleukin 1 β and Tumour necrosis factor α in liver and intestine for both control and treatment. However, the upregulated expression of targeted genes except tumour necrosis factor α was found in head kidney of T1. At the end of the study, GIFT Tilapia when infected with *Aeromonas hydrophila* showed an improved immune response in T1 and T2 with lesser signs of infection than Control. The findings of the present study affirmed the importance of biofloc technology in triggering the immunomodulatory response of GIFT Tilapia with its upregulated immune gene expression and its role as an antimicrobial agent against *Aeromonas hydrophila*. This study suggests the adoption of *in-situ* (T1) based biofloc method to obtain better performance of GIFT Tilapia culture.

INTRODUCTION

Tilapia, considered to be a hardy species, is the second most cultured freshwater fish globally. In 2016, the total production of tilapia was roughly about 6.69 million tonnes (FAO, 2018) and is expected to rise to 7.3 million tonnes by the end of 2030

(Bhera *et al.*, 2018). The most distinct characteristic traits of this species include its euryphagic feeding habit, captive breeding potential, tolerance to high stocking density and improved growth performance in various aquaculture systems.

The rapidly rising global population and decline in capture fisheries has accorded greater significance to aquaculture than ever. However, the expansion of aquaculture is limited to land and water utilization which hinders the productivity of aquaculture activities, particularly in Tilapia farming (Brune *et al.*, 2003; Delgado *et al.*, 2020; Piedrahita *et al.*, 2003; Avnimelech *et al.*, 2008). To overcome these bottlenecks, sustainable intensification by the adoption of advanced culture systems and technologies becomes inevitable to improve the production and productivity of the sector.

One of the best benefits is is on the biofloc technology which requires a minimal or zero water exchange and allows stocking of animals at higher densities. Biofloc are conglomerates of algae, bacteria, protozoans, fecal matter and uneaten feed which are held together in a loose matrix by the secretions of filamentous microorganisms or by electrostatic attraction (Hargreaves, 2013). This technology maintains the carbon and nitrogen content in the culture water and uses the dense microbial biomass to strip the ammonia and serves as a nutritional supplement (Schneider *et al.*, 2005). The external addition of carbon sources to the culture water stimulates the growth of heterotrophic bacteria and its uptake of nitrogen by the production of the microbial protein (Avnimelech, 1999) faster than regular nitrification process (Hargreaves, 2006). The nutrient profile of biofloc ranges from

25 to 50 percent of protein and 0.5 to 15 percent of fat on a dry-weight basis. Bioflocs are also a valuable source of limiting amino acids such as methionine and lysine, vitamins (Vitamin C in the range of 0 to 54 ìg/g dry matter) and limiting mineral such as phosphorus (Crab, 2010). In aquafeeds, dried biofloc can be used possibly to replace fishmeal or soybean meal as cheaper sources of protein. Extensive and traditional systems with no or little use of fishmeal supplies nutrient-rich materials to the culture water enhancing the growth of algae and other indigenous organisms on which the fish can feed (Naylor *et al.*, 2000). The effluent waters from aquaculture systems are used for *ex-situ* biofloc production in suspended growth bioreactors. The biofloc produced can be dried and used as a feed supplement for shrimp or fish (Kuhn *et al.*, 2012). But the uptake of biofloc as feed depends on the nature of species, its feeding ability, size of the animal, and size and density of the floc (Avnimelech *et al.*, 2009).

According to the findings from the previous study, the uptake of biofloc as an additional protein source by freshwater prawn, shrimp, and tilapia indicates that the technology can be applied to both freshwater and seawater culture (Crab, 2010; Crab *et al.*, 2009; Crab *et al.*, 2010a). Biofloc helps in the potential feed gain with decreased production cost (Craig *et al.*, 2002) which can be estimated to be in the order of 10–20% (Schryver *et al.*, 2008). As biofloc technology deals with bacteria and bacterial products, it is likely to come across immunostimulatory compounds exhibiting possible probiotic effects. However, the relative efficiency of *in-situ* and *ex-situ* biofloc with respect to the immune gene expression of the animal has not been attempted so far particularly in GIFT Tilapia. The Genetically Improved

Farm Tilapia (GIFT) strain has been developed using eight different species of Tilapia under selective breeding by World Fish Centre (Eknath *et al.*, 1998) as a consequence to the emergence of new diseases and lack of fish seed availability. The objective of this study is thus aimed to determine the intake of biofloc by GIFT Tilapia using different incorporation methods and its impact on animal immunological performance along with its gene expression.

METHODOLOGY

Experimental Design

A 180-days culture was carried out in the Advanced Research Farm Facility, Madhavaram in Chennai (13.1478° N, 80.2310° E). The experimental group included *in-situ* biofloc developed within the culture systems - Treatment-1 (T1), biofloc incorporated fish feed developed by *ex-situ* method as Treatment-2 (T2), and animals reared without biofloc as control (C). Animals weighing 5.1 g (±0.05) were stocked at a density of $15/m^3$ in lined ponds of 300 m^2 , in all the experimental groups in triplicates. The animals were fed with isoenergetic and isonitrogeneous diet as per their average body weight in all the treatments. The proximate composition of the biofloc and the experimental diets are presented in Table 1 and 2.

Production of Biofloc

In T1, development and maintenance of biofloc in the freshwater culture ponds was adopted as suggested by Taw (2006) at C:N

Table 1: Proximate composition of biofloc

Nutritional Parameters	Composition (%)
Crude Protein	29.82 ±0.60
Crude Lipid	4.45 ±0.5
Crude Fibre	3.51±0.32
Ash	33.2 ±0.7
Acid insoluble ash	11.25 ±0.5
Moisture	8.34 ±0.64
Organic matter	66.8 ±0.32
Total NFE	26.35 ±0.58
Gross Energy Kcal / 100g	331.42 ±5.5

Organic matter (OM) = 100- Ash

Nitrogen free extract (NFE) = 100- (CP+ CL+ CF+ Ash + Moisture)

Gross energy (GE) = (CP X 5.6) + (CL X 9.44) + (CF X 4.1) + (NFE X 4.1) K Cal /100g

Table 2: Formulation and proximate composition of the diets used in the experiments (% dry matter)

Ingredients (%)	Control (C)	*In-situ* (T1)	*Ex-situ* (T2)
Soybean meal	43.55	43.55	33.55
Corn	15.99	15.99	09.10
Fish meal	10.00	10.00	0.00
Biofloc meal	0.00	0.00	26.89
Ricebran	10.00	10.00	10.00
Bentonite	8.54	8.54	8.54
Limestone	4.57	4.57	4.57
Dicalcium phosphate	4.65	4.65	4.65
Cellulose	0.40	0.40	0.40
Sodium chloride	0.50	0.50	0.50
Vitamin & mineral supplemental mix	0.40	0.40	0.40

[Table Contd.

Contd. Table]

Ingredients (%)	**Control (C)**	***In-situ* (T1)**	***Ex-situ* (T2)**
L-Lysine	0.95	0.95	0.95
DL-Methionine	0.35	0.35	0.35
Vitamin –C	0.07	0.07	0.07
BHT(Butylated Hydroxy toluene)	0.02	0.02	0.02
Dry matter	92.34	92.34	92.79
Digestible dry matter (%)	56.45	56.45	55.13
Crude protein (%)	30.15	30.15	30.10
Digestible protein (%)	27.65	27.65	27.11
Gross energy (KJ/g)	14.36	14.36	14.53
Digestible energy (KJ/g)	11.49	11.49	11.78
Ether extract (%)	2.01	2.01	2.04
Ash (%)	19.74	19.74	19.96

ratio of 10:1. The addition of carbon source to maintain the C:N ratio was followed using the method of Avnimelech (Avnimelch *et al.*, 2009) for the transition of the heterotrophic system. For T2, biofloc production was carried out in two indoor raceway tanks (50 tonnes; 15m x 3m x 1m) in six batches at 10-days interval during January to March, 2018. Tanks were filled with used culture water taken from the fish ponds and 100L biofloc inoculum with bacterial floc developed in a separate tank was added to each raceway. Spentwash obtained from M/s. Rajshree Biosolutions Private Ltd was used as a carbon source. The addition of carbon source promotes the heterotrophic bacteria to reduce the organic matter and assimilate the nitrogen waste into microbial protein. The C:N ratio was maintained at

10:1 for the development of biofloc and addition of urea for nitrogen source. Spentwash as carbon source was added for the maximum utilization of left over inorganic nitrogen and to reduce the chance of occurrence of inorganic nitrogen in the form of total ammonia nitrogen in the collected biofloc. On the 7^{th}day, biofloc was collected using harvest pit by closing the aeration and subsequently harvested by passing water in a nylon filter bag with 10 μm pore size. The collected floc was centrifuged at 2000 rpm and the supernatant water was discarded. To remove the traces of ammonia nitrogen level, biofloсs were washed twice with filtered freshwater. Flocs were dried in a hot air oven at 45 °C. The dried flocs were ground into fine powder (less than 200 ìm), packed in airtight containers and kept in refrigerator until experimental diets were made.

Experimental Diets used in the Trial

A diet without biofloc used in C and T1 was compared against the biofloc incorporated diet in T2 by manipulating soyabean meal, cornmeal and fish meal levels. All the ingredients except biofloc powder, amino acids, butylated hydroxyl toluene (BHT) and vitamin-mineral mixture were mixed with water to make dough. The dough was steam cooked using a pressure cooker for 20 min at 15 psi. Bioflocs and other additives were mixed after cooling and the dough was pressed through a pelletizer with 2 mm die and then dried at 60°C till the desired moisture level was reached. The feed was then stored at 4°C until use.

Water Quality Parameters

Temperature (mercury thermometer) and pH (Labtronics) were monitored daily. Dissolved oxygen, BOD, Free Carbon dioxide, Alkalinity, Calcium and Magnesium ion concentration were measured on weekly basis. Nitrate-N (NO_3–N), Nitrite-N (NO_2–N) and Ammonia were (NH_3–N) estimated using the filtered water samples (APHA, 2006) on a weekly basis.

Immunological Parameters and Antioxidant Indicators

Fish were anesthetized to collect blood samples from the caudal vein. EDTA coated vials were used to collect the blood, and to separate the serum, the blood was allowed to clot and centrifuged. Respiratory burst activity was analyzed using the modified method of Anderson and Siwiki (Anderson, 1995). Myeloperoxidase activity in serum was performed according to Quade and Roth (Quade *et al.*, 1997) with slight modifications. The serum sample was analyzed for glucose level using a kit from Beacon diagnostics Pvt. Ltd. The protein estimation of fish serum was carried out by Lowry's method (Lowry *et al.*, 1951). Catalase stress enzyme assay and Superoxide Dismutase (SOD) assay were performed by following the method of Takahara *et al.* (Takahara *et al.*, 1960) and Misra and Fridovich (1972). All these analyses were performed at the end of the experiment.

Growth Parameters

The growth parameters of GIFT Tilapia were monitored on a weekly basis and various growth indices were calculated:

- Weight gain (WG in g) = Final weight- Initial weight
- Feed conversion ratio = Feed given /Body weight gain
- Specific growth rate (%) =Ln (Final weight) –Ln (Initial weight) x 100 /Number of days
- Survival rate (%) =Total number of Fish harvested/Total number of Fish stocked x 100

Gene Expression

The Immune-related gene expression was studied in Head kidney, liver, hepatopancreas and intestine of the experimental animals in all the treatments. The tissue sample was homogenized in TRI Reagent for RNA isolation and the isolated RNA was stored in -20°C for further use. The RNA isolated was converted to cDNA for Metallothionein gene, Cathepsin L, Toll like receptor 7, Interleukin 1 β and Tumour necrosis factor α using the primers listed in Table 3. The cDNA obtained through reverse transcriptase PCR was serially diluted and used for amplification, melt curve analysis and relative quantification of the target genes was carried out using the Real-Time PCR (Applied Biosystem's Real-Time PCR system StepOnePlus®). The temperature cycling parameters for the two-step PCR reaction were as follows: Initial denaturation at 95°C for 10 min, denaturation at 95°C for 15 sec, annealing and extension at 60 °C for 1min for 45 cycles. The PCR was performed with 20 μL total reaction volume containing 10 μL of 2X SYBR®Greenq PCR master mix (Bio-Rad, USA), 1 μL each of forward and reverse primers (10 pmol), 1 μL of template DNA (30–60 ng) and 7 μL of Nuclease free water. The samples were analysed in triplicates and the relative expression was determined by the comparative threshold cycle method 2^{DDCT} (Delta-Delta CT method) using b-actin as internal control (Pfaffl, 2001).

Table 3: Primers used for five immune- related genes in qRT-PCR

S. No	Gene name	Accession	Primers Number	Base pair
1	Metallothionein gene	XM_003447045.5	GCCACTCCTACACCGTCATTC (FP) CTGGCGTTGCTCTTGTCTCTT (RP)	63
2	Cathepsin L	XM_003444107.5	TGTCTTGCTCGTGGGCTATG (FP) CAGCTATTTTTCACCAGCCAGTAG (RP)	63
3	Toll like receptor 7	XM_019352834.2	CCTATTTTGGCAACTGGCATCT (FP) CACTTCACTCCCATTGTTGATCTC (RP)	78
4	Interleukin 1 β	KF747686.1	TGTCGCTCTGGGCATCAA (FP) GGCTTGTCGTCATCCTTGTGA (RP)	63
5	Tumour necrosis factor α	XM_003438427.5	GCTACGACTCCCAGCACTTTG (FP) GCGGTACTGCTCGGATCTCT (RP)	72

Histopathology

The animals were stocked at 1.25 kg/m^3 in the 2000 L FRP tanks in triplicates from all the experimental groups for the challenge study. Before the challenge study, the Lethal dose (LD_{50}) has been derived based on the experiments carried out with four different dosages delivering the bacteria (10^4, 10^5, 10^6 and 10^7). Low relative percent survival was found in tilapias when they were infected with bacteria of 10^7concentrations. The pathogenic dose has been arrived at, based on these results. At the end of the 180-day culture, the experimental animals were challenged with *Aeromonas hydrophila* pathogen obtained from State Referral Laboratory under Tamil Nadu Dr J. Jayalalithaa Fisheries University. The isolate was grown in tryptic soy broth (TSB Hi- Media, India) for 24 h (30-31 °C) and was harvested by centrifugation at 10,000 rpm for 10 min. This was followed by re-suspending the pellet in phosphate buffered saline (PBS, pH 7.2). The suspension in sterile PBS was injected intramuscularly (0.1ml) in healthy tilapia (Yardimci *et al.*, 2011) from all the treatments delivering 10^7 CFU/fish. The infected moribund fish with typical haemorrhagic wounds at the site of injection were sacrificed for the histopathological study after 4 dpi. Kidney, liver, hepatopancreas and intestine were dissected, rinsed in normal saline and fixed in 10% formalin buffer for 24 hrs. After fixation, the tissues were dehydrated in a series of alcohol concentration (70%, 80%, 90%, and 100% respectively), embedded in paraffin, sectioned at 5 mm and later stained with hematoxylin-eosin (H&E) (AlYahya *et al.*, 2018). The histopathological analysis was performed in the Department of Pathology, Madras Veterinary College, Chennai.

Water quality, growth, survival, immunological parameters, antioxidant status and gene expression of the culture animals were analysed using ANOVA to find out any significant difference between the treatments and control and post hoc analysis using Duncan Multiple range test for the significant values. Statistical analysis was performed using SPSS software version 20.0.The significant differences were calculated at 5% level.

RESULTS AND CONCLUSIONS

Water Quality Parameters

The various water quality parameters along with statistical analysis are shown in Table 4.

Different superscripts denote the significant difference (P<0.05) between groups for each parameter.

Temperature found to have no significant difference between the treatments and control. Free CO_2, pH, BOD, Dissolved oxygen, Alkalinity, Nitrate-N, Nitrite-N and Ammonia-N were found to be significantly different between the treatments and control. Calcium and Magnesium ion concentrations were found to be significantly higher in T1 than in control and T2. The floc volume in T1 was maintained at 15 ml/L for the first 60 days of the culture and it was increased to 45 ml/L at the end of the experiment.

Immunological and Antioxidant Indicators

The immunological and antioxidant indicators were analysed and the graphs along with the standard deviation were constructed which are represented in Figure.1.

Table 4: Water quality parameters of experimental groups in the 180 days culture trial of GIFT Tilapia

Parameters	C	T1	T2
pH	7.51 ± 0.01^{a} (7.22-7.4)	7.31 ± 0.02^{b} (7.36-7.66)	7.47 ± 0.01^{c} (7.37-7.75)
Temperature (°C)	30.36 ± 0.29^{a} (28.02-31.4)	30.52 ± 0.32^{a} (28.0-31.3)	30.62 ± 0.30^{a} (28.03-31.3)
DO (mg/l)	6.12 ± 0.05^{a} (4.12-6.34)	5.36 ± 0.04^{b} (3.29-5.78)	5.87 ± 0.04^{c} (3.17-5.92)
Free carbon di oxide (mg/l)	5.82 ± 0.58^{a} (4.06-8.4)	6.65 ± 0.82^{b} (4.15-8.73)	6.04 ± 0.78^{c} (5.06-8.53)
Alkalinity (mg/l)	70.58 ± 0.61^{a} (45.13-81.6)	65.08 ± 0.60^{b} (45.86-84.3)	67.71 ± 0.75^{c} (54.03-86.45)
Calcium ions (mg/l)	54.48 ± 0.57^{a} (50.53-63.41)	57.70 ± 0.58^{b} (49.6-65.73)	55.14 ± 0.66^{a} (47.5-60.24)
Magnesium ions (mg/l)	46.01 ± 0.61^{a} (30.63-62.83)	49 ± 0.61^{b} (31.7-67.1)	45.23 ± 0.62^{a} (32.6-64.2)
Nitrate (mg/l)	0.163 ± 0.0004^{a} (0.001-0.17)	0.124 ± 0.0004^{b} (0.002-0.18)	0.174 ± 0.0004^{c} (0.001-0.18)
Nitrite (mg/l)	0.017 ± 0.001^{a} (0.002-0.02)	0.004 ± 0.0004^{b} (0.002-0.01)	0.007 ± 0.002^{c} (0.002-0.01)
Ammonia (mg/l)	0.154 ± 0.0002^{a} (0.001-0.16)	0.073 ± 0.0003^{b} (0.001-0.08)	0.120 ± 0.0004^{c} (0.001-0.21)
BOD (mg/l)	6.30 ± 0.39^{a} (3.53-8.73)	6.85 ± 0.76^{b} (4.05-8.03)	6.57 ± 0.65^{c} (5.4-7.46)

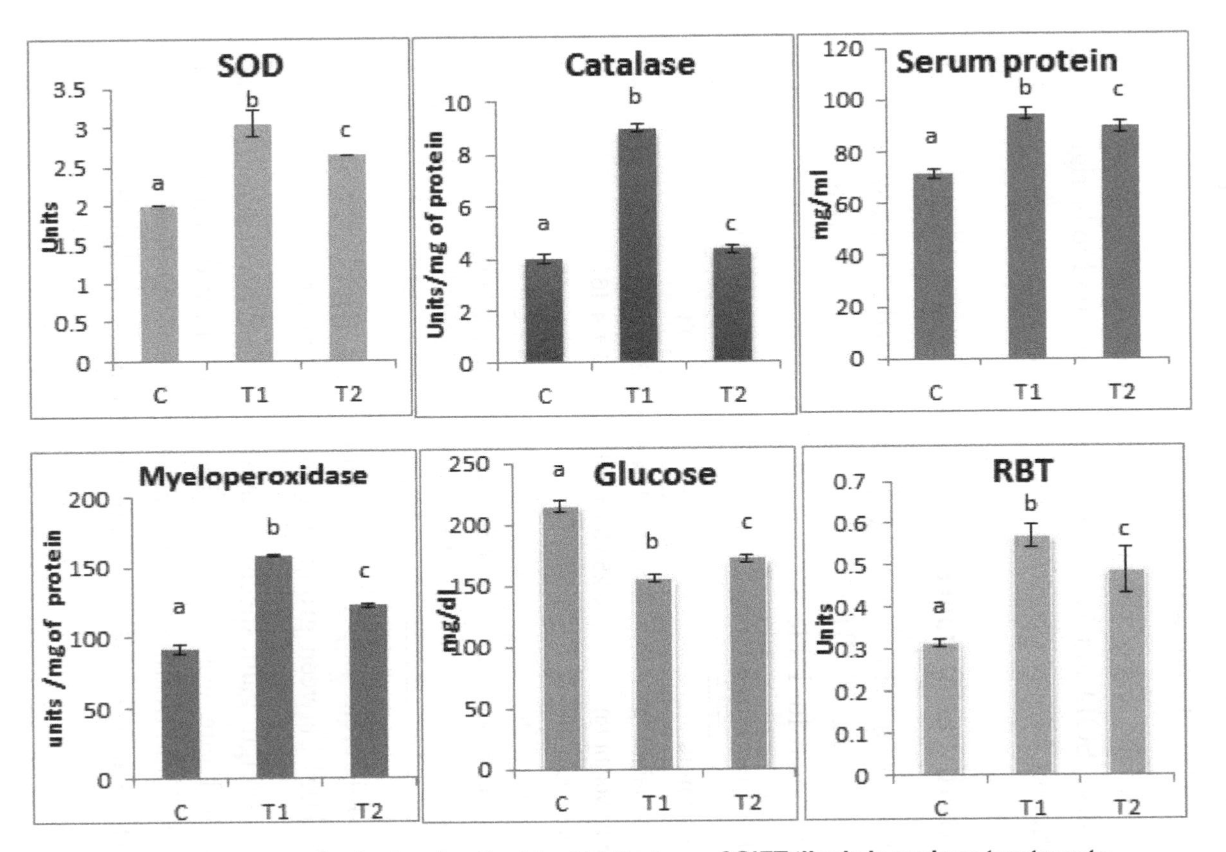

Figure 1. Immunological and antioxidant indicators of GIFT tilapia in various treatments.

Different superscripts denote the significant difference ($P<0.05$) between groups for each parameter.

At the end of the study, serum protein, RBT, glucose levels, catalase, SOD and Myeloperoxidase were found to be significantly different between control and treatments.

Growth Performance

The weight gain, specific growth rate, feed conversion ratio and survival rate of GIFT Tilapia along with the statistical analysis are shown in Table 5.

Table 5: Growth Performance of GIFT Tilapia at the end of the culture trial

Parameters	C	T1	T2
Initial Weight (g)	5.12 ± 0.04^{a}	5.23 ± 0.05^{a}	5.18 ± 0.04^{a}
Final Weight (g)	253.33 ± 4.4^{a}	323 ± 4.16^{b}	282.33 ± 4.33^{c}
Weight gain (g)	248.21 ± 4.39^{a}	317.77 ± 4.12^{b}	277.15 ± 4.3^{c}
Specific growth rate	2.16 ± 0.37^{a}	2.29 ± 0.01^{b}	2.22 ± 0.03^{c}
Feed conversion ratio	1.42 ± 0.01^{a}	1.27 ± 0.01^{b}	1.31 ± 0.005^{c}
Survival rate	83 ± 1.85^{a}	91 ± 1.52^{b}	89 ± 1.15^{c}

Different superscripts denote the significant difference ($P<0.05$) between groups for each parameter.

Weight gain, specific growth rate, feed conversion ratio and survival rate were found to be significantly different between control and treatments. The results of the study showed improved performance of GIFT Tilapia in T1 compared to T2.

Gene Expression

The results of the gene expression showed upregulated immune gene expression in head kidney compared to liver and intestine in all the experimental groups. The gene expression in the head kidney was found to be significantly different between the treatments and control with a higher level of expression in T1. In head kidney, relative mRNA expression of target genes was upregulated except tumor necrosis factor alpha gene. Metallothionein is expressed threefold in T2 whereas in T1, a sevenfold higher expression of this gene was observed. Cathepsin L is expressed fourfold in T2 and sixfold in T1 respectively. Toll like receptor expression levels was up-regulated in both T1 and T2. Interleukin 1 beta gene expressions levels were one to threefold higher in T1 and T2 compared to C. Tumor necrosis factor Alpha gene showed no marked level of expression in all the experimental groups. In liver and intestine there was no folded expression of targeted genes in both control and treatment. The gene expression levels in head kidney are shown in Figure 2.

Different superscripts denote the significant difference ($P<0.05$) between groups for each parameter.

Histopathology

No mortality was observed when the cultured animals were challenged with *Aeromonas hydrophila* at the end of the trial. The results from histopathology showed the presence of lower degree levels of infection in T1 followed by T2 and C. The histopathological analysis of intestine, liver, hepatopancreas and kidney were shown in the Figures 3, 4 and 5.

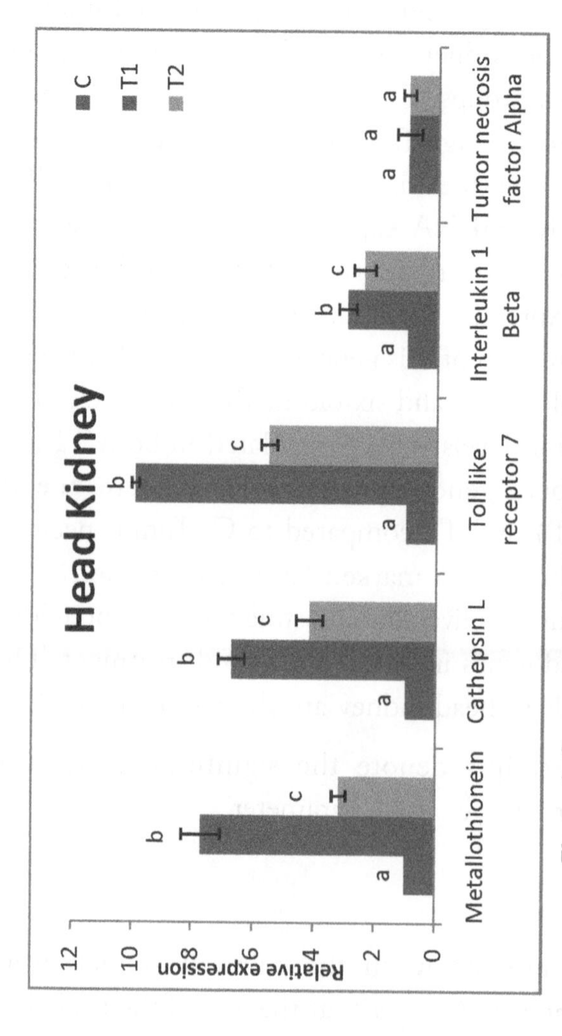

Figure 2. Gene expression levels in the head kidney of GIFT Tilapia in experimental groups

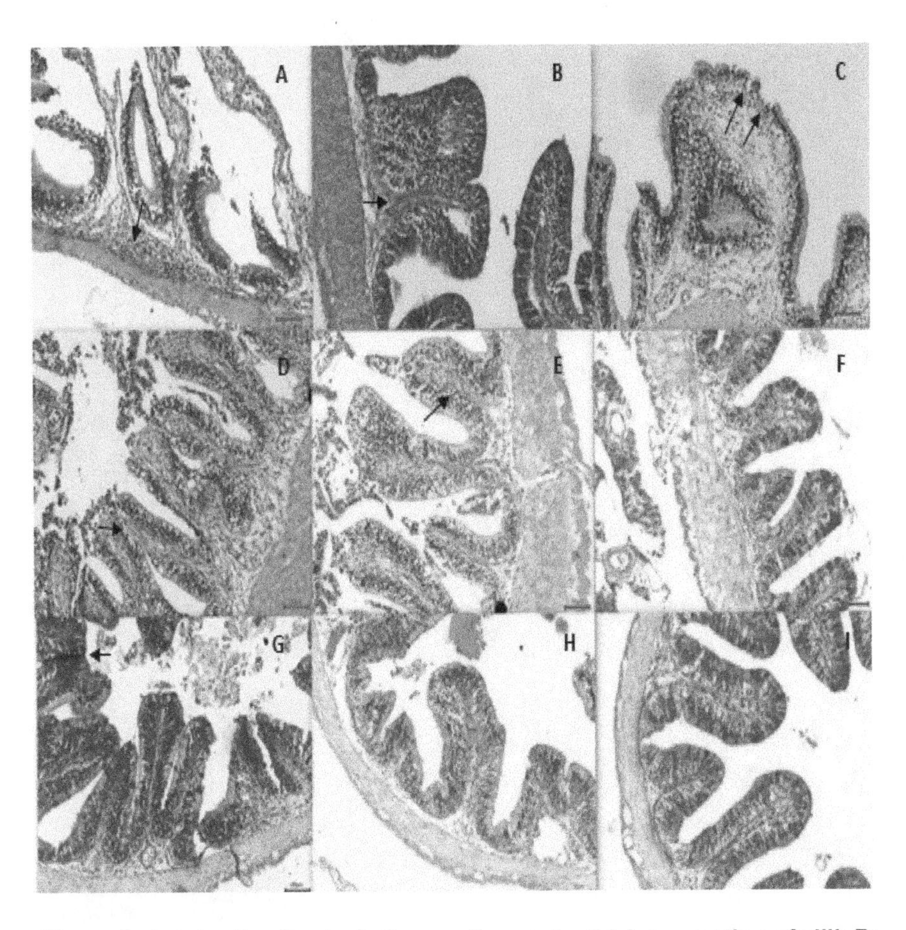

Figure 3: A: Intestine Control - Congestion and mild degeneration of villi; B: Intestine control- Fusion of villi and the separation of lamella propria from the epithelium; C: Intestine control- Mild degenerative necrosis of mucosoepithelial cells; D: T1 Intestine- Mild Inflammation of infiltration cells; E: T1-Intestine- Mild Infiltration of Inflammatory Cells; F: T1 Intestine- NAD; G: T2 Intestine- Fusion of villi and mild degeneration of mucosal epithelium; H:T2 Intestine- Mild inflammation of infiltration cells; and I: T2 Intestine- NAD

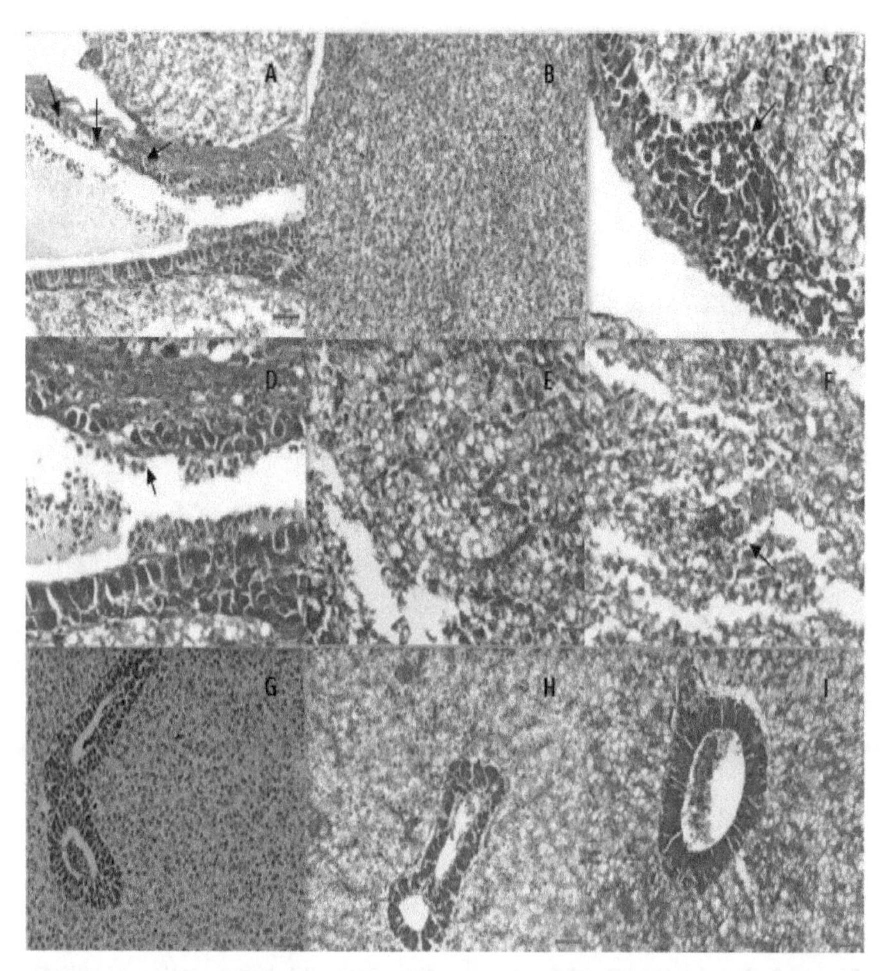

Figure 4: A: Liver Control - Degenerative Necrosis & Congestion of haemorrhages; B: Liver control- Fatty Degeneration of hepatocytes; C: Liver control- Mild haemocytic infiltration & degenerative haemorrhages; D: Hepatopancreas Control - Degenerative Pancreatic Cell Haemorrhages; E: Hepatopancreas Control - Degenerative Sinusoidal Congestion ; F: Hepatopancreas Control -Mild haemocytic infiltration; G: T2 Liver- Mild degenerative changes of hepatocytes; H: T2 Liver- Sinusoidal congestion & mild fatty degeneration and I: T2 Liver - Very mild degeneration of pancreatic cells.

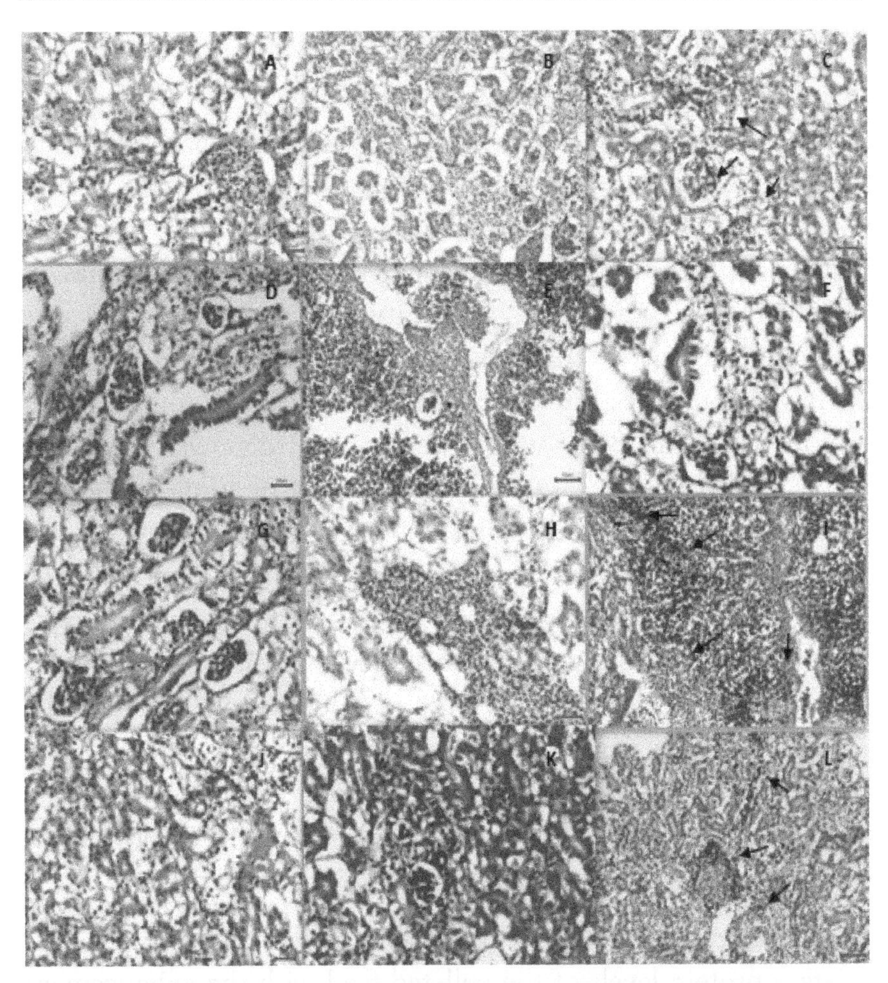

Figure 5: A: Kidney Control - Congestion and vacuolar degeneration of nephritic tubules; B: Kidney Control - Degenerative necrosis tubular epithelial cells; C: Kidney Control - Hyperemia of glomeruli; D: Kidney Control - Mild dilatation of bowman's capsule; E: Kidney Control -Melanomacrophage aggregation and infiltration; F: Kidney Control -Necrosis of tubular epithelial cells along with pyknotic nuclei; G: Kidney Control - Partial loss of glomeruli tuft; H: Kidney Control - Haemorrhages; I: Kidney Control - Melanomacrophage centres and congestion; J: T1 Kidney - Mild tubular degeneration of epithelial cells; K: T2 Kidney- Mild Degenerative tubular epithelial cells ; L: T2 Kidney- Few Melanomacrophage centre aggregation

Temperature and DO (> 3mg/L) in the experimental groups were maintained at levels ideal for the growth of GIFT tilapia (Santos *et al.*, 2013). Lower levels of alkalinity were found in T1 due to the presence of dominant heterotrophic bacterial groups which are responsible for nitrogen uptake due to carbon supplementation. This was in agreement with the studies of Ebeling *et al.*, 2006. As alkalinity concentration alters the buffering capacity of the water it was found that in T1 the effect of low alkalinity leads to lower pH levels. A higher concentration of free CO_2 and BOD with lower levels of dissolved oxygen in T1 was also found. This may be due to respiration by the fish as well as microbes present in the biofloc. However lower levels of CO_2 and BOD were found in control due to its photosynthetic oxygen production. The levels of Calcium and Magnesium were found to be improved in T1 as this ionic concentration influences the floc formation (Schryver *et al.*, 2008) and adhesion by neutralizing the negative charges of the particles. Ammonia-N in T1 remained stable (< 0.02mg/L) throughout the culture trial. The increased levels of Nitrate-N and Nitrite-N in control and T2 indicate the existence of autotrophic nitrification.

The higher level of serum protein in T1 helps to reduce the dietary protein levels of the pelleted feed with the enhancement of the non-specific immune response (Rao *et al.*, 2006). In this study, the RBT of tilapia showed an improved performance in T1 than C and T2. This may be related to the intake of biofloc by the culture animals in T1, which not only boosts the nutrition of the animal but also stimulates the fish cellular defence mechanism in the mode of respiratory burst and phagocytosis (Sharp *et al.*, 1993 and Xu, 2014). The myeloperoxidase (MPO), an antimicrobial enzyme acts by utilizing one of the oxidative radicals to produce hypochlorous acid. The increased MPO activity was seen more in T1 than the other experimental groups. This was concurrent with the findings of Long *et al.*, 2015 who reported increased MPO activity in GIFT when grown in biofloc system for a period of 8 weeks. Increased glycogenolysis and the glucose synthesis from extrahepatic tissue proteins and amino acids aggravates the glucose content in blood as an indicator of stress in animals (Almeida *et al.*, 2001). In the present study, T1 was found to have lesser glucose level when compared with other treatments which in turn indicates the reduced stress level in animals. Biofloc reduced the physiological stress in GIFT which agrees with the studies of Verma *et al.*, 2016 who reported the reduced levels of Cortisol and glucose in *Labeo rohita* when reared in biofloc systems.

The results from the present study revealed increased SOD and catalase level in T1, followed by T2 and C. A spurt in the levels of SOD and catalase improves the antioxidant status of the animal by preventing lipid peroxidation through conversion of superoxide anion to water and oxygen (Tao *et al.*, 2013). Similar studies were done by Yilmaz (Yilmaz, 2019) where Nile tilapia when fed with 5 g/ kg of caffeic acid as a dietary supplement for 60 days improved the fish immune parameters, antioxidant status, as well as survival rate against *A. veronii*. SOD and catalase under hypoxia conditions are involved in the antioxidant defence system by removing and detoxifying oxygen radicals generated within the cells under normal or stressful conditions (Kurata *et al.*, 1993). Lower levels of SOD and catalase indicate cell damage due to the accumulation of high-level free radicals in cells affecting the quality and palatability of fish which impacts human consumption. GIFT Tilapia in T1 & T2 reared under biofloc technology showed improved antioxidant status with increased SOD and catalase levels thus paving the way for easy consumer acceptance. Animals in T1 were found to have increased weight gain, specific growth rate, survival and decreased feed conversion ratio. This may be due to the consumption of microbial floc which is produced as cellular protein by the assimilation of waste nitrogen in the culture animal. The increased intake of the animal in the culture ponds is attributed to the enhanced floc production by the heterotrophic bacterial population in T1 (Wang *et al.*, 2015). The feed response of biofloc incorporated diet in T2 and control was similar as animals tend to jump to fetch feed at the time of application. The animal's response in T1 was not high and this may be due to the existence of biofloc in the culture system consistently throughout the experiment. The total feed applied in T2 and control disappeared in a short span of time, whereas increased feed retention was observed in T1. This led to decreased pellet feeding to the animals in T1. These observations are similar to the findings of Avnimelech (Hargreaves, 2006) as tilapia has the ability to harvest the flocs continuously for feeding in the culture ponds with decreased pellet feeding.

The up-regulation of IL- 1β was observed in head kidney, which indicates its influence in stimulation of immune response. This was also proven from the studies of Kheti *et al.*, 2017 who reported that microbial floc supplemented in the diet of rohu potentiates the expression of IL- 1β and TNF-α in head kidney and liver. Similar kind of upregulated expression of IL-1β and

TNF-α in intestinal tissue was found when *Echinacea purpurea* extract and/or vitamin C in combination or individually supplemented along with the basal diet by Rahman *et al.*, 2018. IL-1β activates the lymphocytes and stimulates the release of other cytokines during the microbial invasion or when there is a tissue injury (Low *et al.*, 2003). TNFs play a role in inflammatory response, proliferation and differentiation of cells, and stimulation of the immune system (Savan *et al.*, 2004 and Wang *et al.*, 2013). The pattern of this cytokine gene expression predicts the changes in immune response. The upregulated expression of these immune genes in T1 enhances the immune cell secretions such as proinflammatory cytokines like TNF-α and IL-1β to modulate the innate immune response of the culture animals. However, there are not too many previous studies reporting the immune gene expression in Tilapia either by rearing in biofloc based culture system or feeding with biofloc meal.

Histopathological manifestations in kidney, liver, pancreas and intestine of GIFT Tilapia against its challenge with *Aeromonas hydrophila* were similar to the observations of Roberts (Roberts, 2012). Degenerative necrosis of tubular epithelial cells along with the melanomacrophage centre aggregation was the major histopathological observation in the kidney. Fatty degeneration of hepatocytes with sinusoidal congestion was found in the liver and pancreas. Fusion of villi with inflammation of infiltration cells and infiltration of inflammatory cells were commonly seen in intestine. These major manifestations were observed with the higher degree of infection in control followed by T2 and T1. This may be due to the toxins and extracellular products produced by *A.hydrophila* such as hemolysin, protease, and elastase causing severe necrosis in the liver and other tissues (Afifi *et al.*, 2000). The infection in T1 fish was found to be lesser due to the production of immunostimulatory compounds (Sakai, 1999 and Abraham *et al.*, 2007) by the heterotrophic bacteria in the biofloc produced within the culture ponds. Microbial floc has also been reported for the presence of bioactive compounds such as carotenoids, polysaccharides, phytosterols, taurine and poly-β-hydroxybutyrate (PHB) (De Schryver *et al.*, 2008; Ray *et al.*, 2010 and De Schryver *et al.*, 2010). The results of the present study can be related to the antioxidant status of the animal and it is found that animals in T1 had a higher immune potential towards the infection followed by T2 and Control. Similar study was performed by Kheti *et al.*, 2017 who administered the microbial floc in the diets of Rohu and showed the increased survival rate when infected with *Edwardsiella tarda*.

Thus, from the above research findings, the present study reveals the improved performance of *in-situ* based biofloc compared to *ex-situ* feeding as it exhibits ideal water quality parameters, improved growth performance, modulatory immune response as well as the upregulated expression of genes responsible for immune system and the resistance towards pathogenic infection.

Biofloc technology is one of the advanced culture technologies adopted for tilapia farming due to its innumerable benefits. It serves as feed for the culture animals, improves the biosecurity of the farm with minimal or zero water exchange. This study and its findings are the first to know the effect of biofloc intake relating to the immunological performance of GIFT Tilapia with gene expression. This gives strong insights on the dietary supplementation of biofloc in feed and its development within the culture ponds for the maintenance of the optimum water quality parameters, growth performance and immune gene regulations in the grow out culture systems of GIFT Tilapia.

ACKNOWLEDGEMENT

This research work was part of the Project funded by the Department of Biotechnology, Govt of India (Project code: DBT-507: PI: S.Felix).

REFERENCES

Abraham, T.J., Babu, C.H.S., Mondal, S. and Banerjee, T. 2007. Effects of dietary supplementation of commercial human probiotic and antibiotic on the growth rate and content of intestinal microflora in ornamental fishes, Bangladesh J. Fish. Res. 11: 57–63.

Afifi, S.H., Al-Thobiati, S., Hazaa, M.S. *et al.* 2000. Bacteriological and histopathological studies on Aeromonas hydrophila infection of Nile tilapia (Oreochromis niloticus) from fish farms in Saudi Arabia, Assiut. Vet. Med. J. 42: 195–205.

Almeida, J.A., Novelli, E.L.B., Dal-Pai Silva, M. and Alves Jr, R. 2001. Environmental cadmium exposure and metabolic responses of the Nile tilapia Oreochromis niloticus, Environ. Pollut. 114 (2001) 169–175.

AlYahya, S.A., Ameen, F., Al-Niaeem, K.S., Al-Sa'adi, B.A., Hadi, S., and Mostafa, A. 2018. Histopathological studies of experimental Aeromonas hydrophila infection in blue tilapia, Oreochromis aureus, Saudi journal of biological sciences. 25: 182-185.

Anderson, D.P. and Siwicki, A.K. 1995. Basic hematology and serology for fish health programs.

APHA (American Public Health Association). 2006. Standard Methods for the Examination of Water and Wastewater, American Public Health Association. Washington, DC.

Avnimelech, Y. 1999. Carbon/nitrogen ratio as a control element in aquaculture systems, Aquaculture. 176: 227–235.

Avnimelech, Y., Kochba, M. 2009. Evaluation of nitrogen uptake and excretion by tilapia in bio floc tanks, using 15 N tracing, Aquacult. 287: 163-168.

Avnimelech, Y., Verdegem, M.C.J., Kurup, M. and Keshavanath, P. 2008. Sustainable land-based aquaculture: rational utilization of water, land and feed resources, MediterrAquac J. 1: 45-55.

Behera, B.K., Pradhan, P.K., Swaminathan, T.R., Sood, N., Paria, P., Das, A., Verma, D.K., Kumar, R., Yadav, M.K., Dev, A.K., Parida, P.K., Das, B.K., Lal, K.K. and Jena, J.K. 2018. Emergence of Tilapia Lake Virus associated with mortalities of farmed Nile Tilapia Oreochromis niloticus (Linnaeus 1758) in India, Aquaculture. 484: 168-174.

Brune, D.E., Schwartz, G., Eversole, A.G., Collier, J.A., Schwedler, T.E. 2003. Intensification of pond aquaculture and high rate photosynthetic systems,AquacEng. 28: 65-86. http://dx.doi. org/ 10.1016/S0144-8609(03)00025-6.

Crab, R. 2010. Bioflocs technology: an integrated system for the removal of nutrients and simultaneous production of feed in aquaculture, PhD thesis, Ghent University. 178.

Crab, R., Chielens, B., Wille, M., Bossier, P., and Verstraete, W. 2010a. The effect of different carbon sources on the nutritional value of bioflocs, a feed for Macrobrachium rosenbergii post larvae, Aquaculture Research. 41: 559–567.

Crab, R., Kochva, M., Verstraete, W. and Avnimelech, Y. 2009. Bio-flocs technology application in over-wintering of tilapia, Aquaculture Engineering. 40: 105–112.

Craig, S., Helfrich and L.A. 2002. Understanding Fish Nutrition, Feeds and Feeding (Publication 420–256), Virginia Cooperative Extension, Yorktown (Virginia). 4.

De Schryver, P., Crab, R., Defoirdt, T., Boon, N., and Verstraete, W. 2008. The basics of bioflocs technology: the added value for aquaculture, Aquaculture. 277: 125–137.

De Schryver, P., Sinha, A.K., Baruah, K., Verstraete, W., Boon, N., De Boeck, G. and Bossier, P. 2010. Poly-beta-hydroxybutyrate (PHB) increases growth performance and intestinal bacterial range-weighted richness in juvenile European sea bass, Dicentrarchus labrax, Appl. Microbiol. Biotechnol. 86 1535–1541.

Delgado, C.L., Wada, N., Rosegrant, M.W., Meijer, S., Ahmed, M. 2003. Fish to 2020: Supply and demand in changing global markets, Washington, D.C.: International Food Policy Research Institute, World Fish Center Technical Report no 62.

Ebeling, J.M., Timmons, M.B. and Bisogni, J.J. 2006 Engineering analysis of the stoichiometry of photoautotrophic, autotrophic, and heterotrophic removal of ammonia–nitrogen in aquaculture systems, Aquaculture. 257: 346-358.

Eknath, A.E. and Acosta, B.O. 1998. Genetic improvement of farmedtilapia project final report (1988-1997), International Center for Living Aquatic Resources Management, Manila, Philippines, 75.

FAO, 2018. Global aquaculture production, in: FaAOotU, Nations (Ed.).

Hargreaves, J.A. 2006. Photosynthetic suspended-growth systems in aquaculture, Aquaculture Engineering. 34: 344–363.

Hargreaves, J.A. 2013. Biofloc production systems for aquaculture, 1-11.

Kheti, B., Kamilya, D., Choudhury, J., Parhi, J., Debbarma, M., Singh, S.T. 2017. Dietary microbial floc potentiates immune response, immune relevant gene expression and disease resistance in rohu, Labeo rohita (Hamilton, 1822) fingerlings, Aquaculture. 468: 501-507.

Kuhn, D.D., Lawrence, A. 2012. *Ex-situ* biofloc technology. In: Biofloc Technology a practical guide book, 2nd edn., edited by Avnimelech, Y., 217-230, Journal of The World Aquaculture Society. Baton Rouge, Louisiana, USA.

Kurata, M., Suzuki, M. and Agar, N.S. 1993. Antioxidant systems and erythrocyte life-spanin mammals, Comp. Biochem. Physiol. B. 106: 477–487.

Long, L., Yang, J., Li, Y., Guan, C. and Wu, F. 2015. Effect of biofloc technology on growth, digestive enzyme activity, hematology, and immune response of genetically improved farmed tilapia (Oreochromis niloticus), Aquaculture. 448: 135–141.

Low, C., Wadsworth, S., Burrells, C. and Secombes, C.J. 2003. Expression of immune genes in turbot (Scophthalmus maximus) fed a nucleotide-supplemented diet, Aquaculture. 221: 23–40.

Lowry, O.H., Rosebrough, N.J., Farr, A.L. and Randall, R.J. 1960. J. Biol. Chem.193 (1951) 265-275.

Misra, H.P. and Fridovich, I. 1972. The role of superoxide anion in the autoxidation of epinephrine and a simple assay for superoxide dismutase, Journal of Biological chemistry. 247: 3170-3175.

Naylor, R.L., Goldburg, R.J. , Primavera, J.H., Kautsky, N., Beveridge, M.C.M., Clay, J. Folke, C., Lubchenco, J., Mooney, H., and

Troell, M. 2000. Effect of aquaculture on world fish supplies, Nature. 405: 1017–1024.

Pfaffl, M.W. 2001. A new mathematical model for relative quantification in real-time RT–PCR, Nucleic acids research. 29: 45.

Quade, M.J. and Roth, J.A. 1997. A rapid, direct assay to measure degranulation of bovine neutrophil primary granules. Veterinary Immunology and Immunopathology. 58: 239-248.

R.H. Piedrahita, Reducing the potential environmental impact of tank aquaculture effluents through intensification and recirculation, Aquaculture. 226 (2003) 35-44. http://dx.doi.org/10. 1016/S0044-8486(03)00465-4.

Rahman, A.N.A., Khalil, A.A., Abdallah, H.M. and ElHady, M. 2018. The effects of the dietary supplementation of Echinacea purpurea extract and/or vitamin C on the intestinal histomorphology, phagocytic activity, and gene expression of the Nile tilapia, Fish & shellfish immunology. 82: 312-318.

Rao, Y.V., Das, B., Jyotyrmayee, P. and Chakrabarti, R. 2006. Effect of Achyranthes aspera on the immunity and survival of Labeorohita infected with Aeromonas hydrophila, Fish Shellfish Immunol. 20: 263-273.

Ray, A.J., Seaborn, G., Leffler, J.W., Wilde, S.B., Lawson, A. and Browdy, C.L. 2010. Characterization of microbial communities in minimal-exchange, intensive aquaculture systems and the effects of suspended solids management, Aquaculture. 310: 130–138.

Roberts, R.J. 2012. Fish pathology, (Fourth ed.), Wiley-Blackwell, UK, 590.https://doi.org/10.1002/9781118222942.

Sakai, M. 1999. Current research status of fish immunostimulants, Aquaculture. 172: 63–92.

Santos, V.B., Mareco, E.A., Silva, M.D.P. 2013. Growth curves of Nile tilapia (Oreochromis niloticus) strains cultivated at different temperatures, Acta Sci. Anim. Sci. 35: 235–242.

Savan, R. and Sakai, M. 2004. Presence of multiple isoforms of TNF alpha in carp (Cyprinus carpio L): genomic and expression analysis, Fish Shellfish Immunol. 17: 87–94.

Schneider, O., Sereti, V., Eding, E.H. and Verreth, J.A.J. 2005. Analysis of nutrient flows in integrated intensive aquaculture systems, Aquaculture Engineering. 32: 379–401.

Sharp, G.J.E. and Secombes, C.J. 1993. The role of reactive oxygen species in the killing of the bacterial fish pathogen *Aeromonas salmonicida* by rainbow trout macrophages, Fish& Shellfish Immunology. 3: 119-129.

Takahara, S., Hamilton, H.B., Neel, J.V., Kobara, T.Y. and Y. Ogura E.T. 1960. Nishimura, Hypocatalasemia: a new genetic carrier state, The Journal of Clinical Investigation. 39: 610-619.

Tao, Y., Pan, L., Zhang, H. and Tian, S. 2013. Assessment of the toxicity of organochlorine pesticide endosulfanin clams Ruditapes philippinarum, Ecotoxicology and environmental safety. 93: 22-30.

Taw, N. 2006. Shrimp production in ASP system, CP Indonesia: Development of the technology from R&D to commercial production, Aquaculture America.

Verma, A.K., Rani, A.B., Rathore, G., Saharan, N., Gora, A.H. 2016. Growth, non-specific immunity and disease resistance of Labeo rohita against Aeromonas hydrophila in biofloc systems using different carbon sources, Aquaculture. 457: 61-67.

Wang, G., Yu, E., Xie, J., Yu, D., Li, Z., Luo, W., Qiu, L. and Zheng, Z. 2015. Effect of C/N ratio on water quality in zero-water exchange tanks and the biofloc supplementation in feed on the growth performance of crucian carp, Carassius auratus, Aquaculture. 443: 98-104.

Wang, T. and Secombes, C.J. 2013. The cytokine networks of adaptive immunity in fish, Fish Shellfish Immunol. 35: 1703–1718.

Xu, W.J. and Pan, L.Q. 2014. Evaluation of dietary protein level on selected parameters of immune and antioxidant systems, and growth performance of juvenile Litopenaeus vannamei reared in zero-water exchange biofloc-based culture tanks, Aquaculture. 426: 181-188.

Yardimci, B. and Aydin, Y. 2011. Pathological findings of experimental Aeromonas hydrophila infection in Nile tilapia (Oreochromis niloticus), Ankara Univ Vet FakDerg. 58: 47-54.

Yilmaz, S. 2019. Effects of dietary caffeic acid supplement on antioxidant, immunological and liver gene expression responses, and resistance of Nile tilapia, Oreochromis niloticus to Aeromonas veronii, Fish & shellfish immunology. 86: 384-392.

CHAPTER 3

FERTILIZATION PROTOTYPE ON BIOFLOC DEVELOPMENT IN GIFT TILAPIA CULTURE

A study was conducted to evaluate the effect of different fertilizers in GIFT Tilapia culture using biofloc technology. Animals (5±0.23g) were stocked at a density of $30m^{-3}$ in 500 litres FRP tanks and spentwash was used as a carbon source to maintain a C:N ratio of 10:1 for 42 days. The experimental group includes fertilization using ammonia sulphate alone (T1) and fertilization using different inorganic fertilizers (T2). No significant differences in FCR, specific growth rate, weight gain and survival of animals were found between the treatments. Proximate composition and fatty acid profile of floc were comparatively rich in T2.Increased solid concentrations with higher Floc volume index and floc sizes were recorded in T2. The rapid floc development along with multiplication of heterotrophic bacteria and decreased vibrio population was observed in T2. The present study confirmed the influence of fertilizers on the physical and nutritional quality of biofloc in GIFT tilapia culture.

INTRODUCTION

The intensification of the aquaculture can be driven by many advanced technologies and one among them was the biofloc technology. This technology allow aquaculture animals to grow at higher stocking densities with minimal water exchange. The nitrogen recovery in the biofloc technology aids in the prevention of disease outbreak by providing proper biosecurity. Unlike recirculatory aquaculture systems this technology does not require any external filtration rather dense microbial biomass, strips the ammonia inturn to serve as a nutritional source. Tilapia is a hardy species and its omnivorous feeding habit accommodates this fish in the intensive culture practices. Its tolerance to the wide environmental conditions has exceeded the production of 5 million tonnes per year with steady growth rate of 5-8 percent globally (Menaga and Fitzsimmons, 2017).The suspended growth in ponds such as phytoplankton, bacteria, live and dead particulates and grazers constitutes the floc. Their growth is influenced by various physico chemical factors such as Temperature, pH, Salinity and aeration of the culture systems. The growth and stability of the flocs formed can also be determined by using different carbon sources and at various C/N ratios. The growth of heterotrophic bacteria can be promoted by the external supplementation of carbon sources such as dextrose, sugar, rice flour, wheat flour, rice bran, molasses etc. However the use of distillery spentwash as a carbon source in the biofloc technology is limited. The synergistic way of utilization of distillery spent wash (DSW) by the microbes for its effective degradation brings an eco-friendly application in aquaculture. The ability of the DSW in rapid release of the carbon content paves the way for the growth of heterotrophic bacteria to assimilate ammonia with decreased concentrations of TSS. Assimilation of ammonium by heterotrophic bacteria takes place rapidly due to the faster growth rate as heterotrophic biomass yield per unit substrate are a factor of 10 higher than autotrophic bacteria (Hargreaves,2006).The use of distillery spentwash as a carbon source has been recommended for shrimp as well as GIFT Tilapia (Menaga *et al.* 2017). The balance of carbon and nitrogen with the external supplementation of carbon will convert the ammonium and other organic nitrogenous waste into bacterial biomass (Schneider *et al.* 2005). The chemo-autotrophic system exhibits the classic increase and sudden fall out in ammonia concentration as the nitrifying bacteria oxidize the nitrite-nitrogen to nitrate-nitrogen with the help of available CO_2 in the systems. Another disadvantage of nitrifying bacteria includes its higher sensitivity to the increased ammonia and decreased dissolved oxygen level which in turn impacts the maintenance of optimum water quality parameters (Masser *et al.* 1999; Villaverde *et al.* 2000; Ling and Chen, 2005). In addition, both the heterotrophic and the chemoautotrophic show very low concentrations of nitrite-nitrogen, because nitrite is not a product in either pathway.The biofloc technology has been implemented in various countries and its adoption strategies vary for different countries. The fertilization prototype adopted by the farmers on biofloc has not been discussed much so far. The use of inorganic fertilizers such as Urea, Triple Super Phosphate has been in practice as a source of Nitrogen and Phosphorus in the aquaculture systems. As the pond fertilization augment the production of plankton thereby the different trophic levels including autotrophic and heterotrophic bacterial population helps in increased fish production (Grag and Bhatnagar, 2000). The objective of the present study was to assess the different methods of fertilization prototype for the biofloc development using distillery spentwash as a carbon source at constant C:N ratio 10:1 in the tank culture of GIFT Tilapia.

METHODOLOGY

The experiment consists of two treatments such as fertilization of water using 20 gm L^{-1} pond soil, 10 mg L^{-1} ammonium sulphate and 200 mg L^{-1} distillery spentwash as described by (Avnimelech and Kochba, 2009) (T1) and fertilization of water using various inorganic fertilizers as suggested by (Taw, 2006) (T2) in freshwater (0 ppt).The list of fertilizers used in T2 study were given in the Table 1. Distillery spent wash (DSW) was used as the carbon source to maintain the C:N ratio at 10:1 and it was added thrice to the culture tanks following standard protocols of Crab *et al.* (2010) with slight modification based on the carbon content of Distillery spentwash. The DSW used in the current study was collected from M/s. Rajshree Biosolutions Private Limited, Coimbatore and stored in room temperature. The main characteristics of DSW were analysed at Central Leather Research Institute, Chennai and the results were listed in Table 2. The GIFT Tilapia fingerlings (5±0.23g) were stocked at a density of $30/m^3$ in $500m^3$ FRP tanks in triplicates. Animals were fed with 30% crude protein feed daily as per their body weight for 42 days. The water quality and floc characteristics were investigated throughout the culture trial.

Table 1: Fertilizers used for biofloc development

Day	Fertilizers	Quantity (g/ton)
1	Urea	1.1
	Triple Super Phosphate (TSP)	0.14
	Grain pellet	4
	Dolomite	7
2	Grain pellet	4
	Dolomite	7
3	Grain pellet	4
	Dolomite	7
4	Grain pellet	4
	DSW	10.1
5	Grain pellet	4
6	DSW	7

Table 2: Physico-chemical properties of Distillery Spentwash

Parameters in mg/L	Results
Colour	Dark Brown Coloured Liquid
pH @ 25°C	6.25
Total Chlorides	7729
Total Suspended Solids	42730
Total Phosphorus	66.6
Total Solids	326980
Calcium	6072.1
Magnesium	4421.4
Potassium	23055
Carbonates	Nil
Electrical Conductivity Micro ohms/cm @ 25 degree	112652
Total Organic Carbon	26%
Total Nitrogen	834

[Table Contd.

Contd. Table]

Parameters in mg/L	Results
Organic Solids	252340
BOD	16500
COD	33520
Bicarbonate	13320
Sulphate	5350
Inorganic Solids	74640

Growth Parameters

The growth parameters of GIFT Tilapia were monitored on weekly basis and various growth indices were calculated:

Feed conversion ratio = Feed given /Body weight

Specific growth rate (%) =Ln (Final weight) –Ln (Initial weight) x 100 /Number of days

Survival rate (%) =Total number of Fish harvested/Total number of Fish stocked x 100

Water Quality Parameters

Temperature (YSI, ProDSS Multiparameter) and pH (Labtronics instruments) were measured daily. Dissolved oxygen, Free Carbon dioxide, Salinity, Alkalinity, Hardness, Calcium and Magnesium ion concentration were measured weekly as per APHA (2008). Water samples were filtered using No.1 Whatman filter paper and collected filtrate was analysed using Resorcinol method for nitrate-N (NO_3 -N) and nitrite-N (NO_2 -N).Phenol hypochlorite method was used for total ammonia nitrogen (TAN) and Orthophosphate (APHA, 2008) were recorded weekly once. Total heterotrophic bacteria (THB) and Vibrio were estimated and expressed as colony forming units (CFU) on weekly basis as per Bergey's manual of systematic bacteriology (Holt *et al.* 1989).

Floc Parameters

Biofloc water was collected using Imhoff Cone and it is kept undisturbed for 20 minutes for the floc to settle down. After 20 minutes the particles settled at the bottom was measured as floc volume (Avnimelech and Kochba, 2009).

Floc porosity was calculated by the volume of water and floc settled in Imhoff Cone according to Smith and Coakley, (1984).

$$\text{Porosity} = (FV/WV)^{*}\ 100$$

Floc volume index was obtained using floc volume and floc concentration (TSS) according to Mohlman (1934). It is calculated using the formula

$$\text{FVI} = \text{Floc volume (ml)/ Floc concentration (g)}$$

Floc density index was calculated using floc volume index and it is the grams of floc which occupies a volume of 100 ml after 30 minutes of settling (WHO international reference centre, 1978).

$$\text{FDI} = 100/\text{FVI}$$

Total organic carbon was analysed according to Walkley and Black (1934). From the titre value TOC was calculated according to the formula,

$$TOC\ \% = [10- (10 * y/x)] * 0.003 * 100/Vs$$

Where x = Titrite value of blank; y = Titrite value of sample; Vs = Volume of sample

Floc size and shape were recorded under digital microscope (Lawrence & Mayo, NLCD-120e). Floc settling velocity was determined using Megara *et al.* (1976). BOD,TS,TDS,TSS and VSS were determined according to APHA (2008) and expressed in mg L^{-1}.These quantitative and qualitative characteristics of floc were determined weekly once during the culture trial.

Biochemical Composition of Biofloc

Proximate composition of the floc were performed as per standard method of AOAC (2005). Extraction of lipid from the biofloc was done as per Folch *et al.* (1957) with slight modifications and fatty acid analysis of the floc was analyzed in GC-MS at Veterinary College and Research Institute, Nammakkal.

Water and Floc quality were analysed using one way ANOVA between treatments at 5% level of significance. Growth parameters of the culture animals were analysed using one way ANOVA and post hoc analysis using Duncan Multiple range test for the significant values. Statistical analysis were performed using SPSS software version 20.0.

RESULTS AND CONCLUSIONS

There was no significant difference in Temperature, Salinity, Dissolved oxygen, pH, Hardness, Magnesium, Ammonia-N, Nitrate-N and Phosphorus. The nitrite concentration pattern was different for both the treatments. Accumulation of nitrite concentrations was seen at the end of the second week in T1 and this may be due to the slow growth of heterotrophs in T1 than T2.The low concentrations of nitrite in T2 reveals the complete assimilation of ammonia to nitrate by heterotrophic bacteria under the similar environmental conditions (Ebeling and Timmons, 2007). In the present study, optimum NH_3-N and NO_3-N concentrations were observed in all the treatments as cited by Hargreaves, 2013 and their accumulated concentrations did not vary with the effect of carbon supplementation. The concentration of CO_2 was relatively higher in T2 (4-4.9 ppm) and this may be due to the ammonia regeneration by the bacterial biomass under biofloc culture systems (Avnimelech, 2015). Total alkalinity and Calcium concentration found significantly different between the treatments. The lower level occurrence in T2 may be due to the buffering action of bacteria as reported by Ebeling *et al.* (2006).

In both treatments the pH and temperature were within the desirable ranges favouring the growth of the culture species (Table 3) (Crab, 2009). The presence of vibrio bacteria population was seen in both the treatments, however inclined multiplication rate was found in T1 compared to T2. This may be due to the faster replication rate of heterotrophs compared to slower growth rate of autotrophs (Ebeling *et al.* 2006).

The mean final weight, survival and FCR values for the fertilization study were presented in Table 5. Based on the statistical analysis; there was no significant differences between the treatments however improved final mean weight of the fishes in T2 culture tanks was recorded. This may be due to the enhanced floc production and thereby the consumption of floc lead to increased weight gain.

Table 3: Water quality parameters of experimental groups for the 42 days of culture trial

Parameters	T1	T2
DO (mg/l)	6.15 ± 0.11^a (4.3 - 6.6)	5.24 ± 0.17^a (5.3 - 6.6)
Temperature (°C)	26.45 ± 0.15^a (25– 27)	26.26 ± 0.1^a (25 – 27)
Salinity (ppm)	0.01 ± 0.001^a	0.01 ± 0.001^a
pH	8.23 ± 0.04^a (8.2 - 8.6)	8.68 ± 0.04^a (8.5 - 8.9)
CO_2 (mg/l)	2.8 ± 0.19^a (1.8 - 3.4)	4.33 ± 0.09^b (4 - 4.9)
Alkalinity (mg/l)	56.77 ± 2.5^a (40 – 63)	47.89 ± 2.9^b (37 – 83)
Hardness (mg/l)	259.2 ± 3.6^a (236 – 292)	249 ± 5.16^a (219 – 279)
Calcium (mg/l)	43.59 ± 4.8^a (46 – 71)	53.1 ± 2.98^b (34 – 72)
Magnesium (mg/l)	79.03 ± 2.9^a (49 – 89)	62.11 ± 2.79^a (41.4 – 92)
NH_3 (mg/l)	0.007 ± 0.0^a (0.001 – 0.044)	0.006 ± 0.0^a (0.004 – 0.011
NO_2 (mg/l)	0.014 ± 0.0^a (0.003 – 0.019)	0.009 ± 0.0^b (0.003 – 0.015)
NO_3 (mg/l)	0.178 ± 0.02^a (0.115 – 0.199)	0.15 ± 0.02^a (0.04– 0.269)
Phosphate (mg/l)	0.082 ± 0.02^a (0.049 – 0.159)	0.09 ± 0.01^a (0.049 – 0.15)

Table 4: Total Heterotrophic bacteria and Vibrio count in culture water of the experimental groups

DOC	TPC (CFU/ml) water			*Vibrio* (CFU/ml) water		
	T1	T2	P value	T1	T2	P value
0	3.11×10^4 $\pm 2 \times 10^3$	5.98×10^4 $\pm 4.3 \times 10^4$	0.001**	6.6×10^2 $\pm 0.04 \times 10^3$	5.1×10^2 $\pm 2 \times 10^1$	0.01*
7	6.37×10^4 $\pm 1.94 \times 10^4$	13.41×10^4 $\pm 2.1 \times 10^4$	0.001**	7.7×10^2 $\pm 0.03 \times 10^3$	4.7×10^2 $\pm 2 \times 10^1$	0.01*
14	12.41×10^4 $\pm 3.2 \times 10^3$	21.63×10^4 $\pm 1.6 \times 10^4$	0.001**	1.01×10^3 $\pm 0.02 \times 10^3$	4.0×10^2 $\pm 2 \times 10^1$	0.01*
21	1.02×10^5 $\pm 2.1 \times 10^3$	3.8×10^5 $\pm 3.8 \times 10^3$	0.01*	1.21×10^3 $\pm 0.01 \times 10^3$	3.1×10^2 $\pm 2 \times 10^1$	0.01*
28	2.34×10^5 $\pm 1.6 \times 10^3$	4.6×10^5 $\pm 2.8 \times 10^3$	0.01*	2.48×10^3 $\pm 0.03 \times 10^3$	2.7×10^2 $\pm 2.1 \times 10^1$	0.001**
35	3.64×10^5 $\pm 2.2 \times 10^3$	5.2×10^5 $\pm 3.1 \times 10^3$	0.01*	4.11×10^2 $\pm 0.01 \times 10^3$	2.1×10^2 $\pm 1 \times 10^1$	0.001**
42	5.9×10^5 $\pm 3.1 \times 10^3$	4.4×10^6 $\pm 4.2 \times 10^3$	0.001**	5.32×10^2 $\pm 0.02 \times 10^3$	1.9×10^2 $\pm 1 \times 10^1$	0.01*

* $P < 0.01$ & ** $P < 0.001$ – significant

Values (Mean ± SE) in the same row with different superscript differ significantly (Duncan Multiple Range Test ($p<0.05$).

Table 5: Growth Parameters of GIFT Tilapia of the experimental groups at the end of the culture trial

Parameters	Treatment 1	Treatment 2
Initial weight (gm)	5 ± 0.23[a]	5 ± 0.20[a]
Final weight (gm)	14 ± 0.16[a]	15 ± 0.12[a]
Weight gain (gm)	8 ± 0.99[a]	9 ± 0.99[a]
Specific growth rate	8.02 ± 0.23[a]	8.16 ± 0.34[a]
Feed Conversion Ratio	1.2 ± 0.04[a]	1.2 ± 0.09[a]
Survival (%)	99 ± 0.02[a]	98 ± 0.4[a]

Values (Mean ± SE) in the same row with different superscript differ significantly (Duncan Multiple Range Test ($p<0.05$).

The small changes in the floc will be highly influenced by the ionic composition of the culture water and its relative changes in the ionic strength will bring a substantial change in the floc morphology. The other factors influencing the physical and chemical characteristics of biofloc includes the type of carbon source, C/N ratio, DO concentrations, shear force caused due to aeration and settling time. The fluctuations of floc quality characteristics throughout the culture trial was given in figure 1 and table 6.

The concentration of Total solids, Total suspended solids and Total dissolved solids found to be higher in T2 than T1.This may be due to the addition of grain pellets and carbon source in the prototype of T2 which paves the way for the faster development of biofloc.Also the percentage of nitrogen in Urea constitutes about 45% and in ammonium sulphate it is closely to 20% (Boyd, 2012). This leads to the presence of increased floc volume in T2 than T1.The inclined trend of BOD level in T2 can be correlated with the abundance of the heterotrophic microorganisms. Significant difference was found between the treatments in total plate count and vibrio count of the culture water (Table 4). The increase in total plate count in T2 with decreased vibrio count revealed the multiplication of beneficial heterotrophic organisms in the culture water. The increased oxygen demand of microbes, culture animal and the organic matter was vividly observed in T2 with the constant C/N ratio under vigorous aeration.

Table 6: Quantitative and Qualitative Characteristics of biofloc in the experimental groups of GIFT Tilapia culture

Parameters	T1	T2
Total Solids (mg/L)	184.4-699	294.4-1240
Total Suspended Solids (mg/L)	98-236	139-589
Total Dissolved Solids (mg/L)	57.9-326	159.4-641
Biochemical Oxygen Demand (mg/L)	3.0-7.4	4.2-8.3
Volatile Suspended Solids (mg/L)	12.2-103.4	48.5-230.9
Total Organic Carbon (%)	0.04-1.54	0.56-1.76
Floc Size (μm)	50.14-378.98	119.54-1058.42
Settling Velocity (mm/sec)	0.47-1.98	0.89-3.15
Porosity (%)	95.11-98.70	97.50-98.74
Floc Volume (ml/L)	0.21-18.76	8-46
Floc Volume Index (ml/g)	0.002-0.079	0.05-0.100
Floc Density Index (g/100ml)	1233.99-2666.67	996.66-1823.07

Mean values in the same row with different superscript differ significantly (P<0.05)

Due to the increased solid concentrations in T2 the presence of higher organic particulate particles was observed with higher Volatile suspended solids. The total organic carbon content of the floc is directly proportional to the VSS and hence the amount

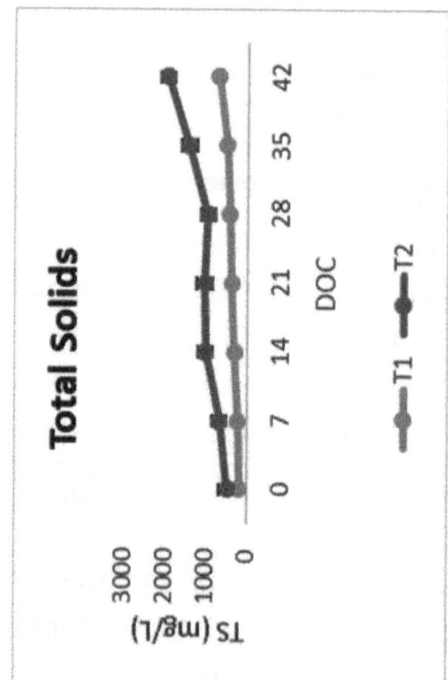

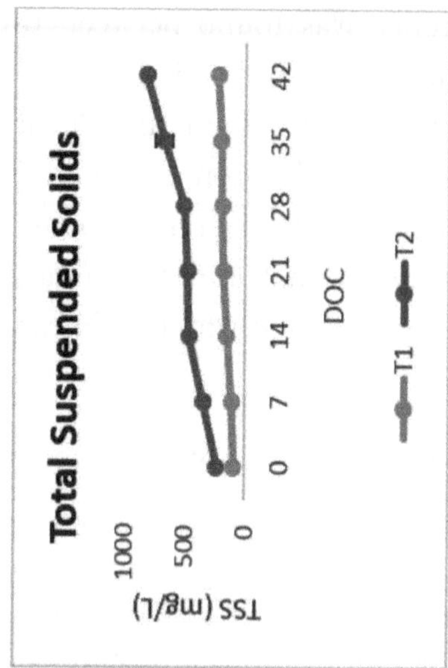

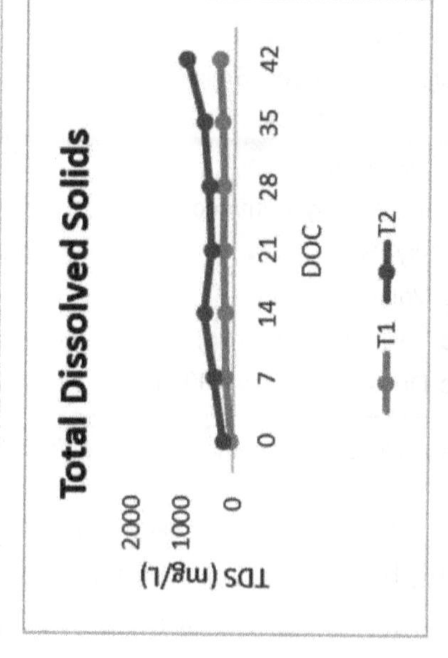

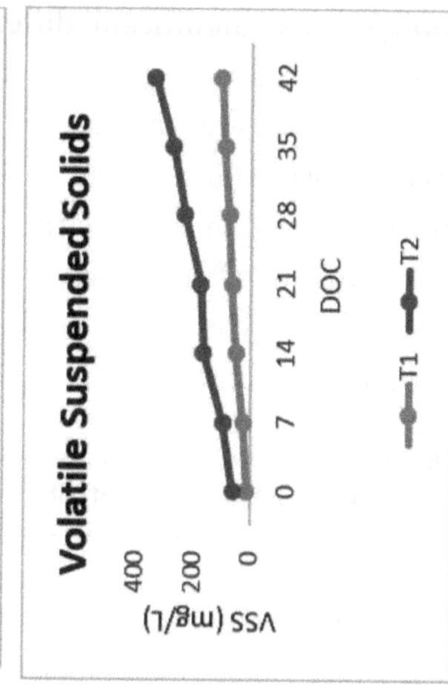

Figure 1: Quantitative and Qualitative Characteristics of biofloc in the experimental groups of GIFT Tilapia culture

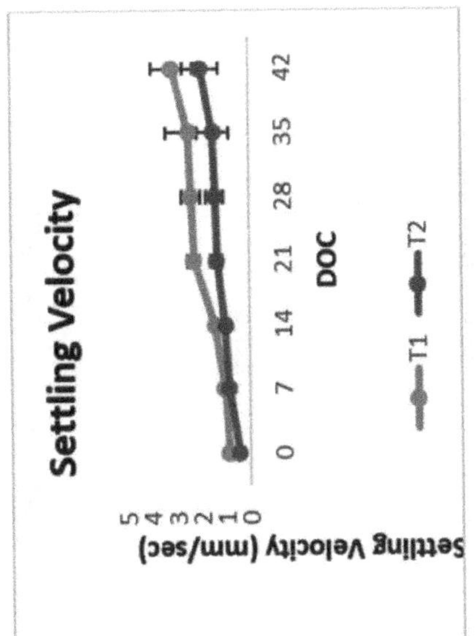

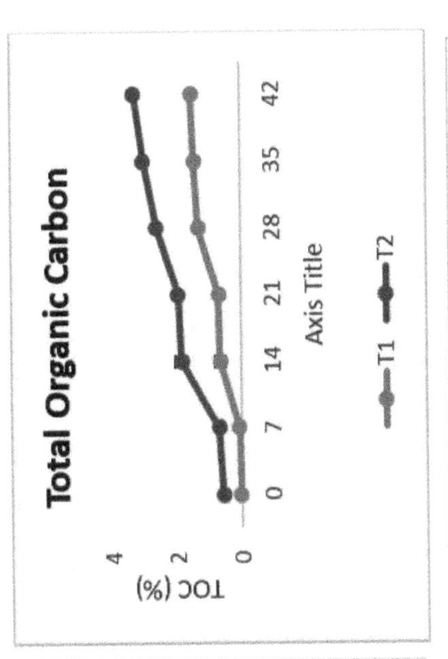

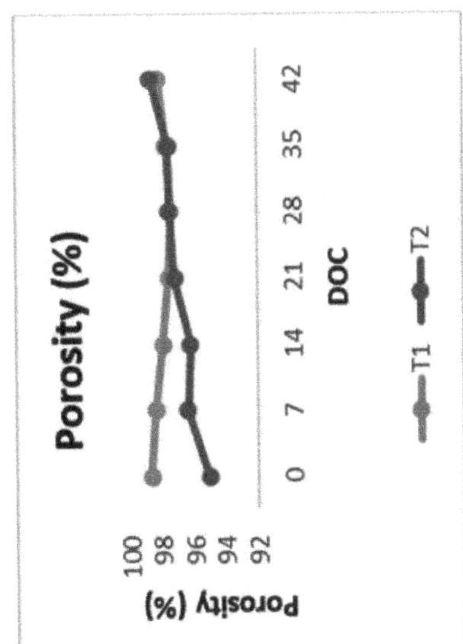

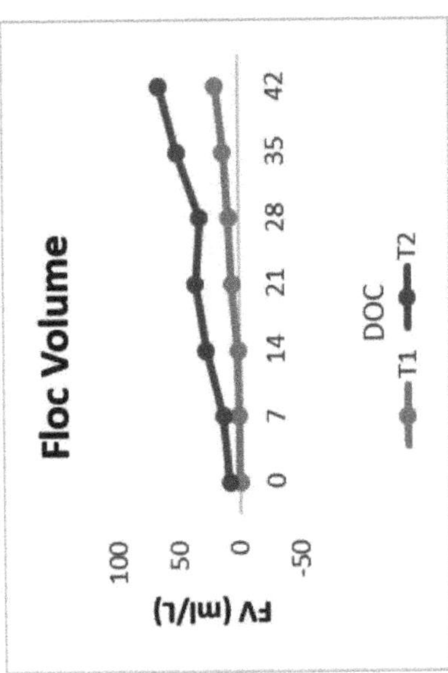

Figure 1: Quantitative and Qualitative Characteristics of biofloc in the experimental groups of GIFT Tilapia culture

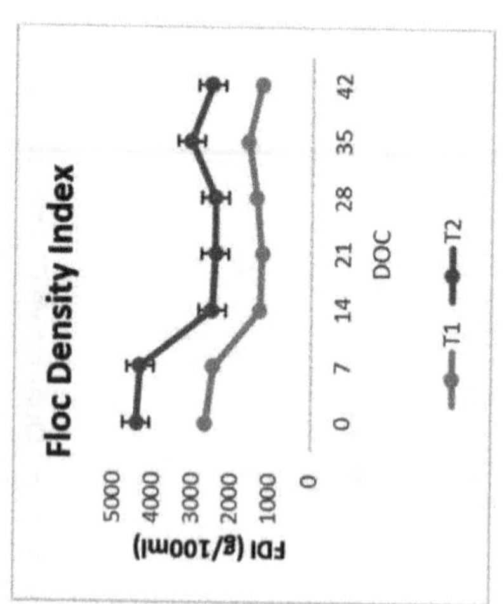

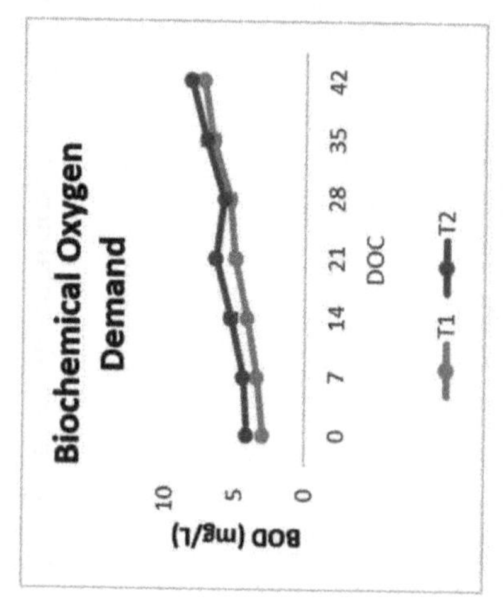

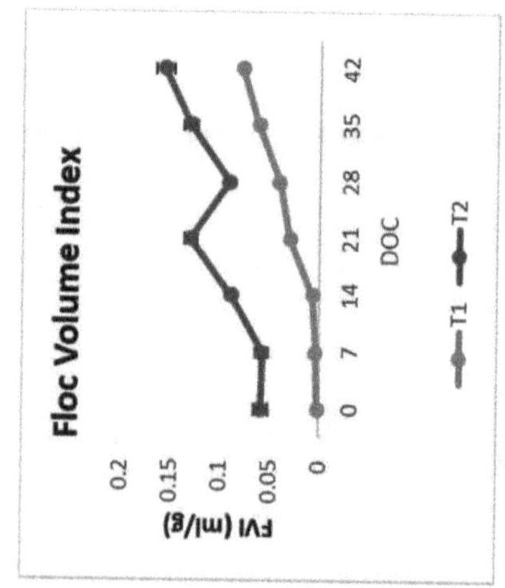

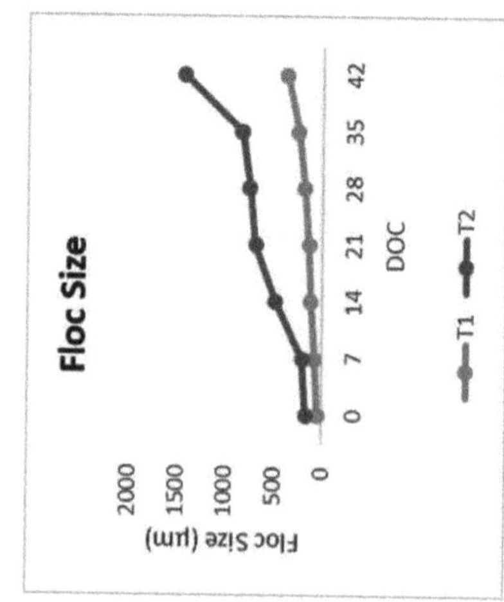

Figure 1: Quantitative and Qualitative Characteristics of biofloc in the experimental groups of GIFT Tilapia culture

of total organic carbon was found higher in T2. The floc size of 250-1200 μm was recommended as it serve as a nutrition source for the aquaculture animals (Barros and Valenti, 2003). The floc sizes observed in the study were within the desirable range for the intake of GIFT Tilapia and a steady state was observed in the treatments based on the floc volume. The floc porosity was higher in T1 compared to T2. The floc porosity is indirectly proportional to the floc size as the flocs are highly porous, the filtration of the suspension will be higher and hence porosity of smaller sized flocs will be higher than larger flocs. The floc volume index (FVI) was found to determine the settleability as well as the age of the floc. Lower the FVI higher the settleablity and it can improve the performance of the culture systems as a source of nutrition thereby maintaining the optimum water quality parameters.

The higher FVI was observed in T2 and this pave more opportunity for the animal to filter the floc with improved floc intake. However the higher FVI also would cause possible clogging of fish gills was also observed as suggested by Avnimelech (2012). The better settleability of floc was observed in T1 due to decreased floc volume and lower FVI. The floc density index (FDI) is an index to determine the compaction and settling ability of floc. The flocs in T1 possessed higher FDI with good compact ability compared to T2.However these changes were observed in a very small level.

In the proximate composition, crude protein, crude lipid and total ash were significantly different ($p<0.05$) between the treatments. This may be due to the rapid formation of floc leading to increased floc volume in T2 by the use of different inorganic fertilizers. These results also indicate the fact that the nutritional quality of floc was not only related to protein content of the feed and carbon source but also on the use of fertilizers triggering the growth of favourable micro-organisms in the culture systems.

The proximate analysis of the biofloc obtained from both treatments were similar to the nutritional profile of floc developed using wheat flour and molasses as carbon sources (Ballester *et al.* 2010). The crude protein content of biofloc collected from two treatments were within the range and confirmed the use of biofloc as a nutritional source to the culture animals (Azim and Little, 2008; De Schryver and Verstraete, 2009; Crab *et al.* 2010; Ekasari *et al.* 2010).

The ash content in the floc may vary based on the accumulation of solid concentrations in the biofloc systems. As the suspended and dissolved solid concentrations were higher in T2 the increased ash content (45%) was found in T2 which is similar to the findings of De Schryver and Verstraete (2009). It is interesting to note the increased gross energy level was observed in the treatments and this may be due to the use of Distillery spentwash as carbon source. The dominant fatty acids such as palmitic acid, linoleic acid and oleic acid in the biofloc samples of T1 and T2 were recorded. The optimum level of Omega 3 fatty acids were observed in T1 and T2 and lower level of PUFA was found in T1 compared to T2.The difference in PUFA level may be due to the huge heterotrophic bacterial population which acts as a nutrient source for the animals to consume (Meyers and Latscha, 1997). The multifold increase of heterotrophic bacteria improved the growth of GIFT Tilapia juveniles in T2 than T1 (Table 7). By the use of different inorganic fertilizers along with dolomite supplementation positively promoted the floc concentration for the maintenance of desirable water quality parameters. In addition the use of grain pellets also triggers the proliferation rate of bacteria for the optimum utilization of nutrients in the biofloc systems. The results of the present study confirmed the use of different inorganic fertilizers as an effective method for the biofloc production in GIFT Tilapia culture tanks.

Table 7: Nutritional Composition of biofloc obtained from the experimental groups in the GIFT Tilapia culture

Nutritional composition	T1	T2
Crude protein	29.82 ± 0.60^{a}	35.22 ± 1.80^{b}
Crude lipid	6.5 ± 0.5^{a}	4.4 ± 0.1^{b}
Crude Fibre	4.71 ± 0.32^{a}	4.6±0.2a
Total Ash	41.2 ± 0.7^{a}	45.5 ± 0.5^{b}
Ether Extract	1.215 ± 0.02^{a}	1.85 ± 0.02^{b}
Total Carbohydrate	23.12 ± 1.31^{a}	25.01 ± 0.7^{b}
Gross Energy KJ g^{-1}	229.5 ± 5.5^{a}	287.7 ± 6.2^{b}
Myristic Acid (14:0)	2.92 ± 0.5^{a}	3.68 ± 0.04^{a}
Palmitic Acid (16:0)	29.69 ± 0.5^{a}	24.995 ± 0.57^{a}
Stearic Acid (18:0)	10.84 ± 0.5^{a}	5.155 ± 0.17^{a}
Oleic Acid (18:1)	31.91 ± 0.5^{a}	21.165 ± 0.98^{b}
Linoleic Acid (18:2n-6)	13.23 ± 0.5^{a}	14.325 ± 1.29^{a}
Linolenic Acid (18:3n-3)	1.5 ± 0.5^{a}	13.05 ± 0.35^{b}
Arachidic Acid (20:1)	0.375 ± 0.5^{a}	1.81 ± 0.19^{a}
Behenic Acid (22:0)	1.24 ± 0.5^{a}	1.445 ± 0.10^{a}
Eicosapentaenoic Acid (20:5n-3)	0.665 ± 0.5^{a}	0.78 ± 0.05^{a}
Docosahexaenoic Acid (22:6n-3)	0.79 ± 0.5^{a}	1.035 ± 0.08^{a}
Palmitoleic Acid (16:1n-7)	6.995 ± 0.5^{a}	9.465 ± 0.51^{b}
Others	0.84 ± 0.5^{a}	0.6 ± 0.01^{a}

The biofloc developed from the different fertilization prototypes influenced the physical and nutritional quality of the floc by favouring the utilization of major nutrients from the fertilizers in the culture tanks. The stability of the flocs and its characteristics vary for the initial days of the culture for the two methods of fertilization however, the steady state floc quality maintenance after thirty days of the culture remained the same. The development of floc was very rapid and increased floc volume in the use of combination of different inorganic fertilizers throughout the experiment was observed. The lower floc volume was maintained in the ammonium sulphate based fertilization in the freshwater. Thus the findings of the study suggested the adoption of use of inorganic fertilizers for the faster development of floc with the predominant heterotrophic bacterial community. The floc stability and floc volume maintenance can be monitored with the addition of the distillery spentwash for the optimum water quality to improve the animal performance.The present findings can be further studied in different salinities as this biofloc technology is applied in both fresh water and sea water.

REFERENCES

AOAC. 2005. Official method of Analysis. 18th Edition, Association of Officiating Analytical Chemists, Washington DC, Method 935.14 and 992.24.

APHA (American Public Health Association). 2008. Standard Methods for the Examination of Water and Wastewater. American Public Health Association, Washington, DC.

Avnimelech, Y. and Kochba, M. 2009. Evaluation of nitrogen uptake and excretion by tilapia in bio floc tanks, using 15 N tracing. *Aquacult.*, 287: 163-168.

Avnimelech, Y. 2015. Biofloc Technology-A Practical Guide Book (3rd Edn.). The World Aquaculture Society, Baton Rouge, United States, pp-37-46

Avnimelech, Y. 2012. Biofloc technology- A Practical Guide Book (2nd Edn.). The World Aquaculture Society, Baton Rouge, Louisian, United States.

Azim, M.E. and Little, D.C. 2008. The biofloc technology (BFT) in indoor tanks:Water quality, biofloc composition, and growth and welfare of Nile tilapia (Oreochromis niloticus). *Aquacult.*, 283: 29-35.

Ballester, E.L.C., Abreau, P.C., Cavalli, R.O., Emerenciano, M., Abreu, L. and Wasielesky, W. 2010. Effect of practical diets with different protein levels on the performance of *Farfantepenaeus paulensis* juveniles nursed in a zero exchange suspended microbial flocs intensive system. *Aquaculture Nutri.*, 16:163-172.

Boyd, C. E., and Tucker, C.S. 2012. Pond aquaculture water quality management. Springer Science & Business Media.

Crab, R., Kochva, M., Verstraete, W. and Avnimelech, Y. 2009. Bio-flocs technology application in over-wintering of tilapia. *Aquacult. Eng.*, 40(3): 105-112.

Crab. R., Chilelens, B., Wille, M., Bossier, P. and Verstraete, W. 2010. The effect of different carbon sources on the nutritional value of bioflocs, a feed for Macrobrachium rosenbergii postlarvae. *Aquacult. Res.*, **41**: 559-567.

de Barros, H. P. and Valenti, W. C. 2003. Food intake of *Macrobrachium rosenbergii* during larval development. *Aquaculture*, **216**(1-4): 165-176.

De Schryver, P. and Verstraete, W. 2009. Nitrogen removal from aquaculture pond water by heterotrophic nitrogen assimilation in lab-scale sequencing batch reactors. *Bioresource Technology*, **100**(3): 1162-1167.

Ebeling, J. M. and Timmons, M. B. 2007. Stoichiometry of ammonia-nitrogen removal in zero-exchange systems. *World aquaculture.*

Ebeling, J.M., Timmons, M.B. and Bisogni, J.J. 2006. Engineering analysis of the stoichiometry of photoautotrophic, autotrophic, and heterotrophic removal of ammonia-nitrogen in aquaculture systems. *Aquacult.*, **257**: 346–358.

Ekasari, J., Crab, R. and Verstraetehayati, W. (2010). Primary nutritional content of bio-flocs cultures with different organic carbon sources and salinity. *HAYATI J. Biosci.*, **17**(3): 125-130.

Folch, J., Lees, M., and Sloane Stanley, G. H. 1957. A simple method for the isolation and purification of total lipides from animal tissues. *J biol Chem*, **226**(1): 497-509.

Grag, S.K. and Bhatnagar, A. 2000. Effect of fertilization frequency on pond productivity and fish biomass in still water ponds stocked with Cirrhinusmrigala (Ham.). *Aquac. Res.*, **31**: 409-414.

Hargreaves, J.A. 2013. Biofloc production systems for aquaculture. *SRAC*, 4503: 1-12.

Hargreaves, J.A. 2006. Photosynthetic suspended-growth systems in aquaculture. *Aquac. Eng*, 34: 344–363.

Holt, J. G., Williams, S. T. and Holt. 1989. Bergey's manual of systematic bacteriology, Vol. 4. Lippincott Williams & Wilkins.

Ling, J. and Chen, S. 2005. Impact of organic carbon on nitrification performance of different biofilters. *Aquac. Eng*, 33: 150–162.

Masser, M.P., Rakocy, J. and Losordo, T.M. 1999. Recirculating aquaculture tank production systems — management of recirculating systems. *SRAC Publication*, 452.

Menaga, M. and Fitzsimmons, K. 2017. Growth of the Tilapia Industry in India. World *Aquaculture.* 49.

Menaga, M., Felix, S. and Gopalakannan, A. 2017. Distillery wastage (spentwash) as a Novel carbon source for Aquaculture Intensification. *Indian Vet. J.*, 94(12): 15 - 17.

Meyers, S.P. and Latscha, T. 1997. Carotenoids. In: D'Abramo, L.R., Conklin, D.E., Akiyama, D.M. (Eds.), Crustacean Nutrition, *Advances in World Aquaculture,* 6. World Aquacult Soc, Baton Rouge, LA, pp. 164–193.

Mohlman, F.W. 1934. The sludge index. *Sewage Works Journal,* 119-122.

Schneider, O., Sereti, V., Eding, E.H. and Verreth, J.A.J. 2005. Analysis of nutrient flows in integrated intensive aquaculture systems. *Aquac. Eng.*, 32: 379–401.

Smith, P.G. and Coackley, P. 1984. Diffusivity, tortuosity and pore structure of activated sludge. *Water Res.*, 18(1): 117-122.

Taw, N. 2006. Shrimp production in ASP system, CP Indonesia: Development of the technology from R&D to commercial production. *Aquaculture America.*

Villaverde, S., Fdz-Polanco, F. and García, P.A. 2000. Nitrifying biofilm acclimation to free ammonia in submerged biofilters, Start-up influence. *Water Res.,* 34: 602–610.

Walkley, A. and Black, I.A. 1934. An examination of the Degtjareff method for determining soil organic matter, and a proposed modification of the chromic acid titration method. *Soil Sci.*, 37(1): 29-38.

WHO international reference center. 1978. Methods of Analysis of Sewage Sludge Solid Waste and Compost, Switzerland, 49.

CHAPTER 4

DISTILLARY SPENT WASH AS A CARBON SOURCE IN INTENSIVE BIOFLOC NURSERY REARING OF *PENAEUS VANNAMEI*

A thirty days experimental trial was performed to investigate the effect of aerobic microbial floc system (AMFS) and aerobic microbial floc/ biofloc incorporated feed (AMFF) on water quality, growth performance, survival and the digestive enzyme activity of the *Penaeus vannamei* in 30-tonnes raceways. Shrimp post larvae (PL15) were stocked at a density of 1500/m^3 in different treatments such as aerobic microbial floc system + commercial feed (AMFS+CF), aerobic microbial floc system + aerobic microbial floc incorporated feed (AMFS+ AMFF), clear water system + commercial feed (CWS + CF) in triplicates. Distillery spent wash was used as a carbon source in the AMFS to maintain the C/N ratio at 15:1. At the end of the experiment, AMFS + AMFF showed significantly ($P < 0.05$) higher digestive enzyme activity such as protease, amylase, and lipase in stomach, intestine and hepatopancreas of the shrimp. AMFS + AMFF showed significantly ($P < 0.05$) higher growth performance, survival (89.44±1.23) and better FCR (1.044±0.002) than the other treatments. Significant difference ($P < 0.05$) in RNA:DNA ratio was found in CWS + CF followed by AMFS. The results from the present study suggests the suitability of AMFS+AMFF treatment for the enhanced growth and digestive enzyme activity of white leg shrimp.

INTRODUCTION

To encounter the growing food demand of the rapidly growing haman population the shrimp farming industry is likely to shift from semi intensive rearing to intensive rearing systems with multiple crops/year. Intensive aquaculture systems efficiently produce dense biomass of shrimp and fish. However, the intensive system results in a rapid accumulation of feed, toxic materials, organic matter and nitrogen compounds in the culture system (Sahu *et al.*, 2013b; Sun and Boyd, 2013). Around 70 to 80% of nitrogen which is added as input to the aquaculture system, remains unutilized and released to the adjacent environment in the form of ammonia, organic nitrogen in feces and residues (Funge-Smith and Briggs, 1998; Avnimelech and Ritvo, 2003; Jackson *et al.*, 2003; Sahu *et al.*, 2013a). Intensive shrimp culture system faces several inevitable environmental problems such as deterioration of water quality, eutrophication, pathogen spread, and disease outbreak (Avnimelech, 2007; Zhang *et al.*, 2012; Liu *et al.*, 2014). To overcome these constraints, nitrogen removal techniques were developed and adopted. Aerobic microbial floc

(Biofloc) has recognized as one among many advanced technologies for solving the impacts aforesaid to attain sustainable aquaculture (Avnimelech, 2007; Crab *et al.*, 2007; De Schryver *et al.*, 2008).

Aerobic microbial floc technology has been gaining importance and found to be lucrative for treating *in-situ* culture water without affecting the shrimp production. This technology mainly depends on the manipulation and regulation of carbon/ nitrogen ratio (C/N ratio) at 10:1 to 20:1 in culture water through the addition of carbon source (molasses, sugar, wheat, feed etc.) and uptake of ammonia by microbial community (Avnimelech *et al.*, 1994; Avnimelech, 1999; Crab *et al.*, 2007) to minimize the regular water exchange. The nutrients from excretion and remnant feed are recycled by microbial community such as heterotrophic bacteria, protozoa, phytoplankton and zooplankton of high quality feed for fish and shrimp (McIntosh, 2000; Avnimelech, 2007) and reduce the potential spread of pathogen (Burford *et al.*, 2003; Crab *et al.*, 2010; Xu and Pan, 2013). Biofloc floc improves the water quality, growth performance, and FCR in a zero water exchange intensive culture system of shrimp (Schveitzer *et al.*, 2013; Xu *et al.*, 2012).

Although microbial floc technology has been applied and developed in intensive shrimp culture system especially for *P. vannamei* (Crab *et al.*, 2007; Ballester *et al.*, 2010; Xu and Pan 2012; Xu *et al.*, 2012), the information concerning how microbial floc can maximize the nutritional benefits to the cultured shrimp is limited. Also, the impact of floc intake directly or through feed by shrimp has not been investigated much in the earlier studies. This study was designed and conducted to evaluate the effect of aerobic microbial floc system (AMFS) and aerobic microbial floc incorporated feed (AMFF) on inorganic nitrogen control, the digestive enzyme activity of the shrimp, growth performance of *P. vannamei* in the intensive nursery culture system

METHODOLOGY

Experimental Design

The experiment was carried out in raceways with 30-ton capacity, over a period of thirty days in indoor nursery raceways at Brackish Water Research Farm Facility, TNJFU-OMR Campus, Fisheries College and Research Institute, Ponneri, TNJFU. A clear water system and aerobic microbial floc system were compared with AMFF and commercial shrimp feed containing a similar level of crude protein (35%). The experimental group includes aerobic microbial floc system + commercial feed (AMFS+CF), aerobic microbial floc system + aerobic microbial floc incorporated feed (AMFS+ AMFF), clear water system + commercial feed (CWS + CF).

Distillery spent wash (DSW) (24% of the carbon source) was collected from M/s. Rajshree Biosolutions Pvt. Ltd. situated at Coimbatore district, Tamil Nadu, India. It was used as a carbon source for maintaining the feed carbon-nitrogen ratio at 15:1, in AMF system (Avnimelech, 1999) and clear water system was operated with regular water exchange throughout the trial.

Seed Stocking and Tank Management

PL 15 shrimp (0.02±0.01 g) (*Penaeus vannamei*), used for the experiments were procured from Aqua Nova shrimp hatchery Kanathur, OMR Road, Chennai and acclimatized for 3 days at 20 ‰. The seed samples were tested for WSSV, AHPND and EHP infections at Molecular diagnostic laboratory, Dr. MGR Fisheries College and Research Institute, Ponneri. The healthy shrimp were stocked in raceways at the rate of 1500 PL/m^3 (Silva *et al.*, 2015) and shrimp were fed with CF and AMFF containing 35% crude protein (Table 1). Feeding was done manually at four times a day (6.00, 10.00, 14.00, and 18.00) based on the average body weight of the shrimp.

Table 1: The proximate composition of the (% dry weight basis) feeds used for culture

Parameters	Commercial feed	Aerobic Microbial Floc incorporated Feed
Moisture	11.66	13.57
Protein	36.35	35
Fiber	2.85	3.62
Fat	5.62	6.77
Total ash	8.99	10.94
Gross energy (kcal kg^{-1})	4083	3885

Water Quality and Biofloc Parameters

The physicochemical parameters of water from the experimental groups has been analyzed daily for the whole experimental period. Visual analysis of water quality, such as color and flocculation was also observed daily. Temperature, salinity, pH, total alkalinity, dissolved oxygen (DO), total hardness, nitrate-nitrogen(N-NO_3), nitrite-nitrogen(N-NO_2), ammonia-nitrogen(N-NH_4^-), total suspended solids (TSS) were measured following standard protocols (APHA, 2005). AMF volume was obtained using Imhoff cones for every three days (Avnimelech and Kochaba, 2009).

Aerobic Microbial Floc and Shrimp Sampling

At the end of experiment, floc samples were collected through 20-micron mesh nylon filter bag. The samples were dried in a hot air oven at 105°C until a constant weight was achieved. The crude protein, lipid, fiber, and carbohydrate of the microbial floc and shrimp were analyzed using standard methods (A.O.A.C., 1995). Protein was determined by measuring nitrogen using the Kjeldahl method multiplying with 6.25; lipid by ether extraction using soxhlet and ash by oven incineration at 550°C, moisture content was determined by hot air oven drying at 105°C for 24 hours.

100 shrimp were collected randomly from each treatment for enzyme analysis at the end of the experiment. Shrimp hepatopancreas, stomach and intestine were pooled, and homogenized.The homogenate (10%) was prepared with phosphate buffer (pH 7.5 and 0.1 M). 0.5 ml of Tween 20 and 1 ml of homogenized sample was taken for the extraction of the enzyme. The suspension was centrifuged at 5000 rpm for 15 minutes and the supernatant was collected.

Enzyme Activity

Amylase activity assay was done according to 3,5 dinitrosalicylic acid colorimetric method by Clark (1964) using starch as the substrate. Protease activity was analysed by azocasein as the substrate (Sarath *et al.*,1989). Lipase activity was determined by Cherry and Crandel, (1932) based on the measurement of fatty acid release due to enzymatic hydrolysis of olive oil. Enzyme activity was measured as the changes in the absorbance, using the spectrophotometer (Lamba 25 UV Win Lab V 6.0) and expressed as specific activity ($Umg^{-1}protein^{-1}{}_{-}min{-}^{-1}$). Total soluble protein content was measured by adopting Bradford (1976) method.

Growth analysis

At the end of the experiment, the growth performance of stocked shrimp was estimated as follows:

i) Percentage weight gain = [Final weight (g) – Initial weight (g) X 100] / Initial weight (g)

   Specific growth rate {SGR (%/day)} = [(ln final weight – ln initial weight) ×100] / Rearing period (days)

ii) Feed Conversion Ratio = Feed given (dry weight) / Body weight gain (wet weight)

iii) Protein Efficiency Ratio = Body weight gain (wet weight) / Crude Protein fed

iv) Survival (%) = (Total number of shrimp harvested / Total number stocked) x 100

Quantitative determination of nucleic acids (RNA-DNA ratio) in shrimp abdomen muscle was done by Pentose analysis following the method of Schneider, (1957).

Statistical analysis were performed using SPSS 24.0 for Windows. One-way analysis of variance (ANOVA) was performed to examine the difference in growth parameters, water quality parameters, digestive enzyme activity and RNA- DNA ratio, among the treatments. The post hoc analysis was performed using Duncan's multiple range test to determine the significant difference among the treatments.

RESULTS AND CONCLUSIONS

Water Quality and AMF Development

Water quality parameters such as TAN, alkalinity, and BOD significantly differed in all treatments (Fig. 1). No significant difference was observed in dissolved oxygen, temperature, salinity, pH, and hardness. The documented water quality in all experimental groups remained within the limit for shrimp culture throughout the culture period (Table 2).

Table 2: Water quality parameters of experimental groups for the 30 days of culture trial

Parameter	AMFS CF	AMFS AMFF	CSW CF
DO (ppm)	5.2 –8.6	5.2 – 7	3.2 – 5.6
Temperature (°C)	25–30	25–30	25–30
Salinity (‰)	19.5–21.5	19.5–21.5	19.5–21.5
pH	7–8.5	7–8.5	7–8.5
Nitrite ($N\text{-}NO_2$) (mg / L^{-1})	0.1–1.2	0.1–1.2	0.1–1.1
Nitrate ($N\text{-}NO_3$)(mg / L^{-1})	0.3–1.0	0.3–1.0	0.6–1.3

Values (Mean ± SE) in the same row with different superscript differ significantly (Duncans Multiple Range Test ($p<0.05$)). All the values in percentage were transformed for ANOVA.

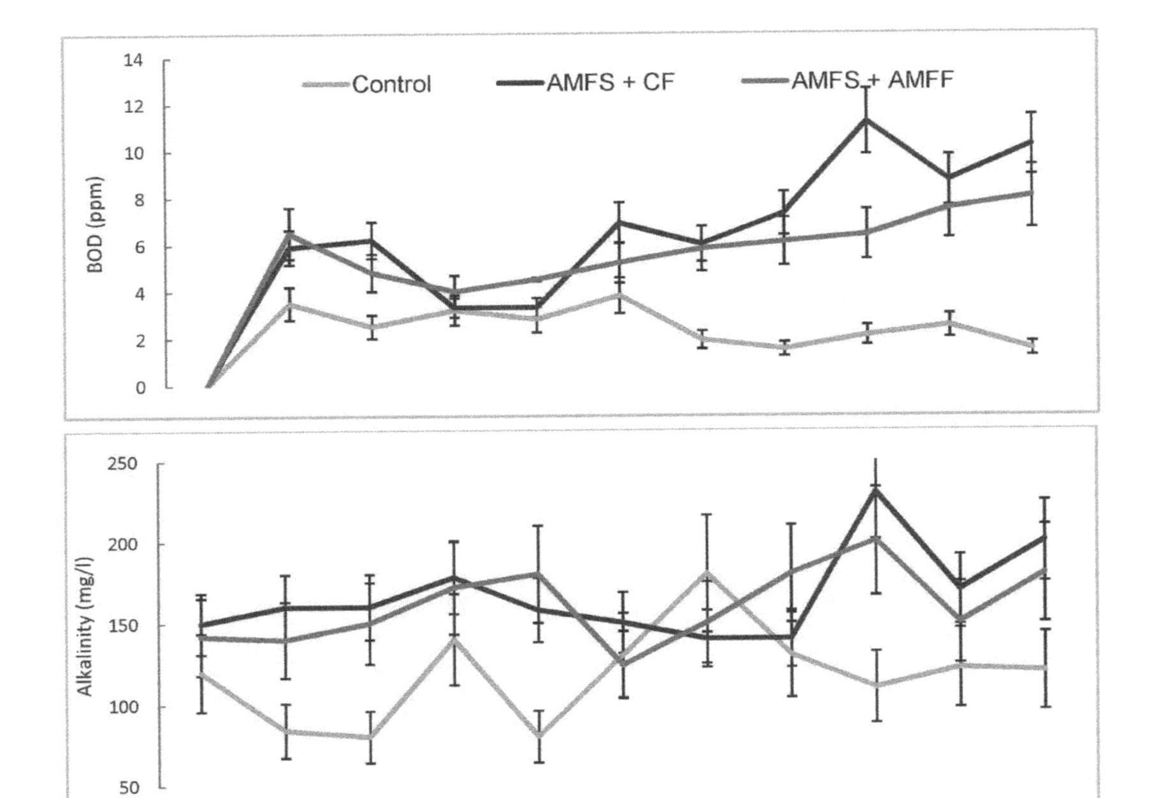

Figure 1. Fluctuation of BOD, Alkalinity and Ammonia concentration in different experimental groups stocked with *P.vannamei* during the 30 days experimental period

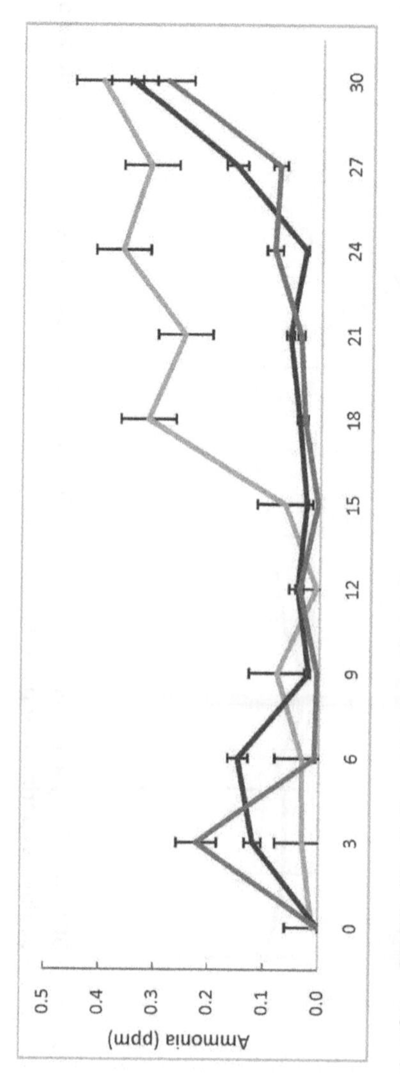

Figure 1. Fluctuation of BOD, Alkalinity and Ammonia concentration in different experimental groups stocked with *P.vannamei* during the 30 days experimental period

Significantly higher TSS, and floc volume was recorded in AMFS+AMFF and FVI was significantly higher in the AMFS + CF treatment. (Fig. 2). The average floc volume and TSS ranged from 14 –18 ml L^{-1} and 300 -- 420 mg L^{-1} in AMFS based treatments.

Digestive Enzyme Activity of Shrimp

Significant differences (P>0.05) in the digestive enzyme activity were observed in all the treatments. The specific activity of the protease, amylase, and lipase in hepatopancreas, stomach, and intestine were significantly higher in AMFS + AMFF followed by CWS + AMFF (Fig 3).

Zoo Technical Performance of the Shrimp

The growth performance of the shrimp in AMFS+AMFF was significantly (P<0.05) higher than other treatments (Table. 3). Mean survival rate was higher than 88% in AMFS based treatments. AMFS+AMFF treatment showed a better FCR and PER when compared to other treatments.

RNA - DNA ratio was observed fortnightly in different experimental groups during the culture period (Fig. 4). The ratio between RNA and DNA ranged from 0.20 to 0.33 in all the experimental groups. At DOC 15, the ratio was higher in AMFS + CF and lowest level was observed in AMFS + AMFF. At the end of the experiment (30th day), CWS + CF reared shrimps showed the highest RNA -DNA ratio, which was significantly different (P<0.05) from other treatment groups.

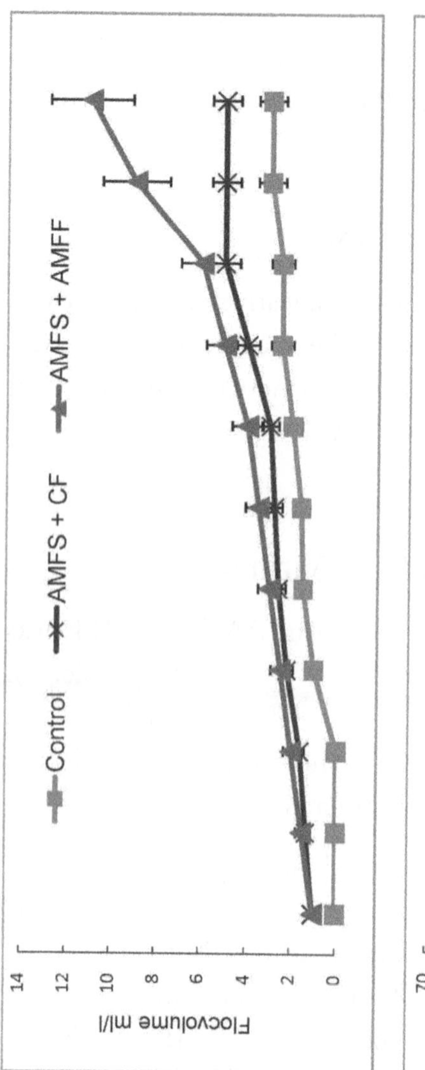

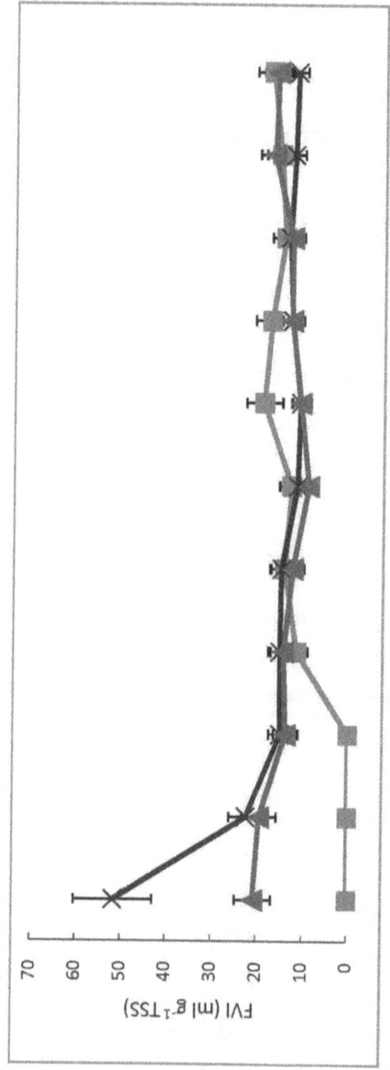

Figure 2. Changes in floc volume, total suspended solids (TSS), and floc volume index (FVI) in different experimental groups stocked with *P.vannamei* during the 30 days experimental period.

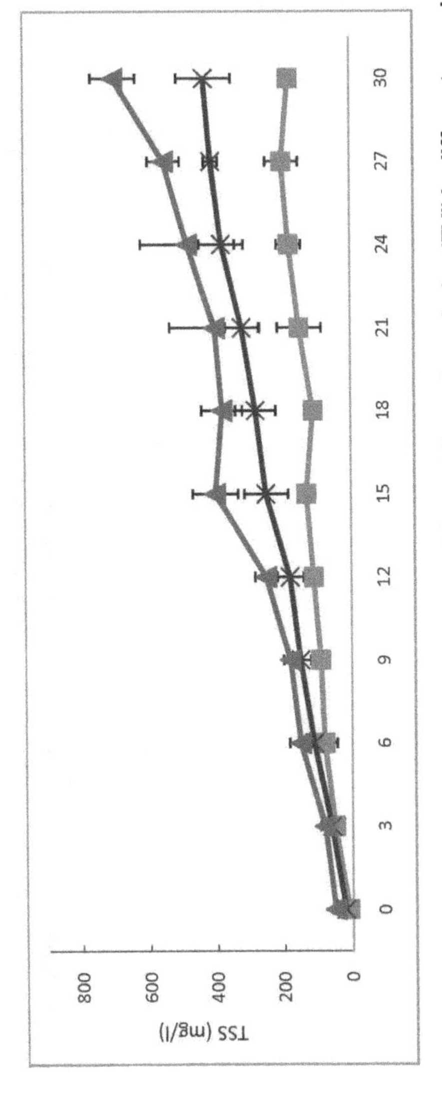

Figure 2. Changes in floc volume, total suspended solids (TSS), and floc volume index (FVI) in different experimental groups stocked with *P.vannamei* during the 30 days experimental period.

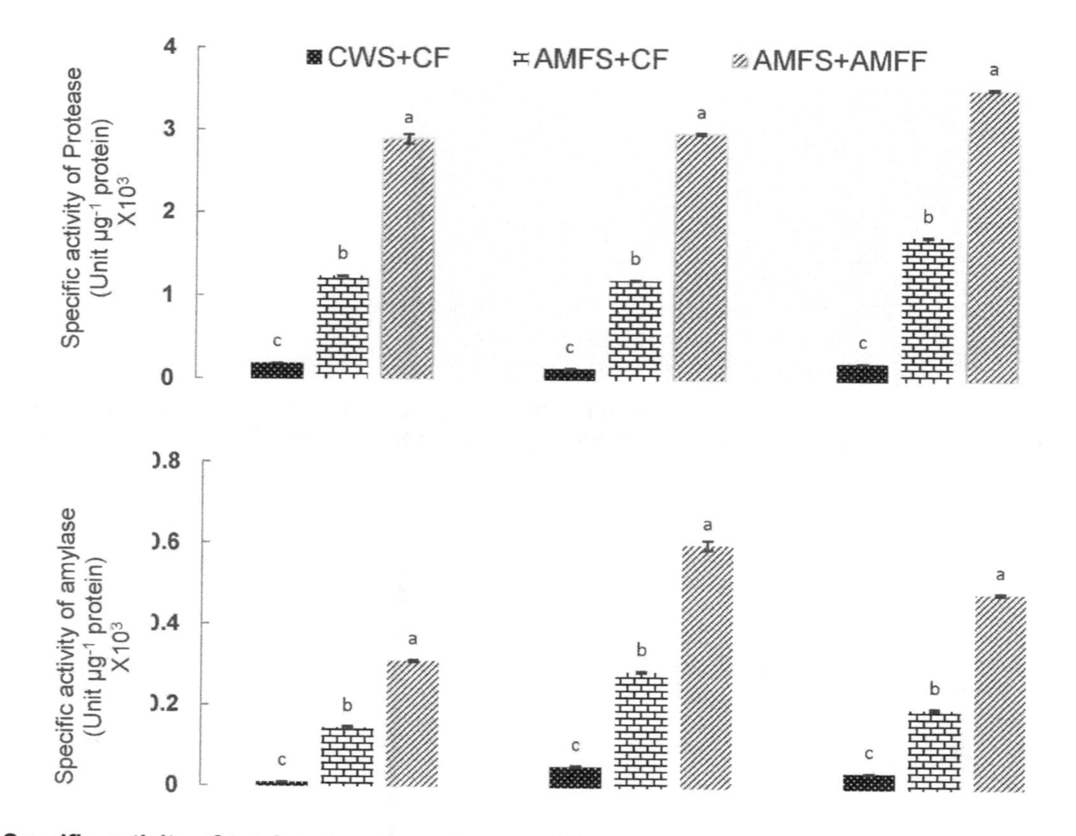

Figure 3. Specific activity of total protease, amylase, and lipase in the hepatopancreas (H), intestine (I), and stomach (S) of *P.vannamei* in different experimental groups at the end of 30 day feeding experiment.

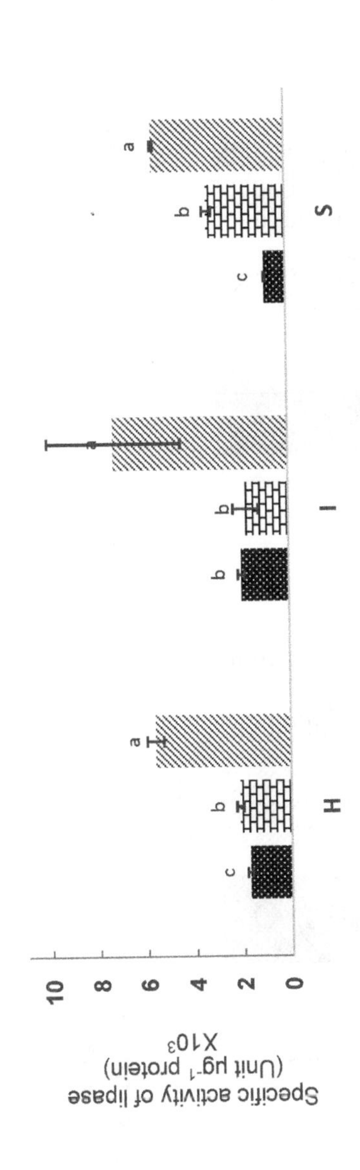

Figure 3. Specific activity of total protease, amylase, and lipase in the hepatopancreas (H), intestine (I), and stomach (S) of *P.vannamei* in different experimental groups at the end of 30 day feeding experiment.

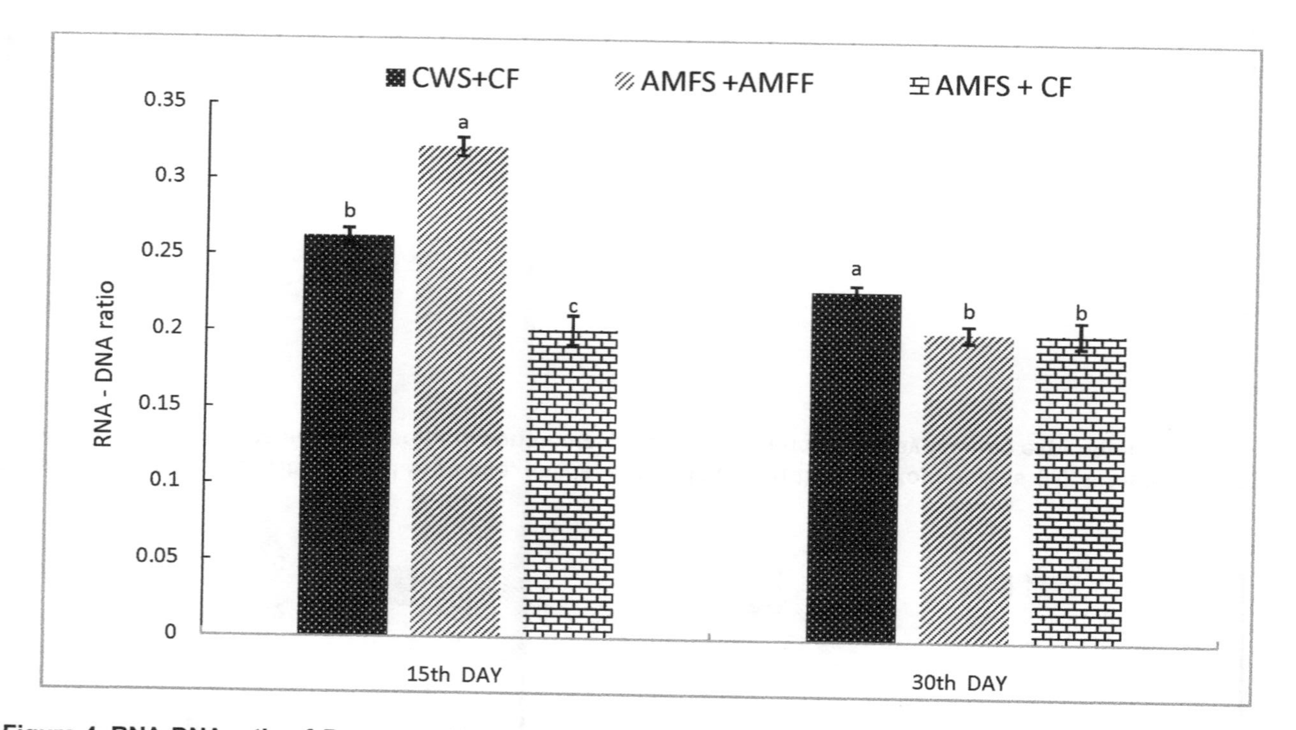

Figure 4. RNA-DNA ratio of *P.vannamei* in different experimental groups at the middle (15 days) and end of (30 days) feeding experiment.

Table 3: Growth performance and feed utilization of juvenile *Penaeus vannamei*

S. No	Growth parameters	CSW +CF	AMFS +CF	AMFS +AMFF
1.	Initial body weight (g)	0.02	0.02	0.02
2.	Final body weight (g)	$1.0048^{b}\pm0.052$	$1.0322^{a}\pm0.01$	$1.0352\ ^{a}\pm0.025$
3.	Percentage weight gain (%)	49.24 ± 0.01^{b}	50.61 ± 0.05^{a}	50.76 ± 0.05^{a}
4.	Survival rate (%)	$79.99^{b}\pm0.5$	$88.88^{a}\pm1.1$	$89.44^{a}\pm1.23$
5.	SGR (% day^{-1})	$13.05^{b}\pm0.2$	$13.14^{a}\pm0.5$	$13.15^{a}\pm0.3$
6.	FCR	$1.08^{b}\pm0.001$	$1.047^{a}\pm0.005$	$1.044^{a}\pm0.002$
7.	PER	$0.028^{b}\pm0.001$	$0.023^{c}\pm0.001$	$0.031\ ^{a}\pm0.001$

Values (Mean ± SE) in the same row with different superscript differ significantly (Duncans Multiple Range Test ($p<0.05$)). All the values were transformed for ANOVA. *SGR: Specific Growth Rate; FCR: Feed Conversion Ratio; FER: Feed Efficiency Ratio; PER: Protein Efficiency Ratio.

Water Quality Parameters

Water quality parameters recorded in all the experimental groups during the experiment were in the optimal range for the nursery rearing of *P. vannamei.* This indicates that the experimental condition was conducive to the growth of the shrimp in AMF (Felix *et al.*, 2015). Dissolved oxygen observed in this study (4 – 8.5 mg L^{-1}) improved the survival and growth of *P. vannamei* (McGraw *et al.*, 2001). The adequate DO level in all the treatments during the experimental period was attributed due to continuous aeration. Water pH in all the treatments was found within the ideal range of 7.0-8.8 (Van Wyk *et al.*, 1999). The alkalinity levels were found optimum as suggested by Boyd *et al.*, 2002 and this might have improved the physiological conditions of shrimp allowing shedding and proper formation of the exoskeleton, promoting growth and survival of organisms (McGraw and Scarpa, 2003).

In the present study, use of distillery spent wash to maintain the C/N ratio 15:1 was efficient to remove all forms of nitrogen compounds (NH_4-N, NO_2-N, NO_3-N) without a substantial increase in BOD (Avnimelech, 1999). DSW assimilated the ammonia and thus produced good quality microbial floc protein (Felix *et al.*, 2015; Liu *et al.*, 2017). The presence of high concentration of NO_2-N & NO_3-N in AMFS treatments indicates the occurrence of nitrification processes in the culture systems. The nitrogen dynamics and reduction in TAN concentration confirms the conversion of organic nitrogen by using distillery spent wash as a substrate and this finding was consistent, with those of the previous studies (Liu *et al.*, 2017; Raj *et al.*, 2017; Felix *et al.*, 2015; Luo *et al.*, 2013; Yao *et al.*, 2013; Lu *et al.*, 2012).

In AMFS, floc volume and TSS levels increased gradually and were kept within the acceptable range for shrimp culture (Samocha *et al.*, 2007). Several authors have indicated a similar trend of concentration of TSS, which is beneficial to the shrimp and the system stability (De Schryver *et al.*, 2008). The levels of TSS in AMFS+AMFF treatment were significantly higher than the AMFS + CF. These differences stemmed due to the ingredient composition of AMFF and intake of floc by the culture shrimp.

Digestive Enzyme Activity

Biochemical composition of the diet plays a vital role in the digestive enzyme profile of shrimp (Gamboa-delgado *et al.*, 2003).

In the present study, dietary supplementation of AMFF significantly (P<0.05) improved the specific activity of the digestive enzymes like protease, amylase, and lipase in AMFS+AMFF treatment compared to other treatments. The increased digestive enzyme activity in AMFS+AMFF treatment enhanced the digestion and nutrient absorption of shrimp and recorded significantly higher (P<0.05) growth rate. A previous study by Xu and Pan, (2012) in microbial floc based system reported similar results in *P. vannamei.* This may also be due to the interference of live microorganisms attached to the aerobic microbial floc after transition through the stomachs with resident intestinal microflora (Moss *et al.*, 2000), which plays an important role in the production or secretion of digestive enzymes (Harris, 1993; Moss *et al.*, 2000; Moss *et al.*, 2001; Xu and Pan, 2012). Thus, the aerobic microbial floc could influence the digestive processes of the shrimp due to the presence of enzymes secreted by microbial floc, inducing endogenous enzymes and altering intestinal microflora balance.

Similarly, an enhanced level of digestive enzyme activity has been reported in fish and shrimp fed with probiotic, microalgae and periphyton supplemented diet (Anand *et al.*, 2013 and Lara-Flores *et al.,* 2003) and the presence of microbial compounds in the microbial floc supplemented diet might have stimulated the production of an endogenous enzyme by the shrimp digestive organs (Ziaei-Nejad *et al.* 2006). Despite the higher ash and fiber level recorded in the AMFF diet, the cultured shrimp showed higher specific enzyme activity compared to other treatments. Nevertheless, this may be because, the shrimp possess a suite of a digestive enzymes capable of hydrolyzing a variety of substances and several factors have been implicating in altering digestive enzyme activity including diet (Lee *et al.*, 1984).

Biogrowth Parameters

In the present study, *P. vannamei* reared under AMFS using distillery spent wash showed a better performance in terms of mean body weight, SGR, FCR, PER, survival, and yield compared to CWS. The results agreed with the findings of Xu and Pan, (2012); Serra *et al.* (2015) as AMFS impacts positively on shrimp growth performance. The inclusion of aerobic microbial floc as a dietary ingredient in shrimp diet (AMFF) found to improve the growth performance and relatively higher survival of *P. vannamei* when compared to CWS+CF treatment (Ju *et al.*, 2008; Kuhn *et al.*, 2009, 2010). This difference was mainly due to the composition of microbial floc, as they are the rich source of many bioactive compounds such as carotenoid, chlorophylls, phytosterols, bromophenols, amino sugars (Ju *et al.*, 2008) and antibacterial compounds (Crab, 2010). The presence of microbial compounds, unknown growth factors, beneficial micro-organisms in the Aerobic microbial floc might have resulted in a significantly higher growth rate and better FCR in shrimp fed with AMFF. FCR in AMFS was better when compared to CWS and agrees with the findings of Avnimelech *et al.* (1994) who found that lowering feed application up to 30 % of conventional feeding ration, did not lower shrimp growth. This may be due to the presence of AMF, where animals were fed continuously with floc and thus supplementary feed intake was less leading to improved FCR.

AMFS treatments resulted in higher survival due to the provision of essential nutrients, such as amino acids, vitamins, and minerals present in the flocs (Decamp *et al.*, 2002).

Quantification of RNA - DNA Ratio in Muscle

The RNA content in AMFS+ CF was high on 15th day compared to other treatments. As the somatic growth of shrimp is directly related to the protein synthesis in tissues. RNA-DNA ratio is directly correlated with protein synthesis as RNA is the precursor for the protein synthesis. RNA-DNA ratio was significantly higher in AMFS than CSW due to favorable water quality parameters and nutrient rich feed availability in the system. It indicates the production of higher level of protein and tissue formation, caused by aerobic microbial floc (Wilder and Stanley 1983). The changes in nucleic acids implicate changes in growth performance of the shrimp and this parameter as an indicator for growth, and nutritional status of the shrimp was observed between AMFS and CWS (Tewary and Patra, 2008).At the end of the trial, the RNA:DNA ratio were found in the decreasing trend in AMFS treatments and this may be due to stress caused by the increased biomass. However, the result obtained from the present study was within the suitable range of 0.20 to 0.32 and agreed with the findings of Moss, (1994).Thus presence of AMF improved the growth performance in AMFS treatments.

Distillery spent wash induced nitrogen management through aerobic microbial floc system will go long way, enabling enhanced shrimp production with reduced input cost. Aerobic microbial floc system could significantly impact the NH_3-N, NO_2-N concentration and TSS, FV level in the system under limited water exchange. All water quality parameters were maintained within the optimal ranges for shrimp culture. The supplied aerobic microbial floc incorporated diet (AMFF) and AMFS positively influenced the growth performance and feed utilization of shrimp by enhancing the digestive enzyme activities. This study, elucidates the suitability of aerobic microbial floc system and incorporation of aerobic microbial floc in shrimp feed for the nursery rearing of *P. vannamei.*

ACKNOWLEDGMENT

The research work was part of the project funded by the Department of Biotechnology, Govt of India , New Delhi (DBT – Project code : 507; PI: Dr.S.Felix).

REFERENCES

A.O.A.C. 1995. Official methods of analysis of AOAC International. *Arlington, Va.: AOAC Intl. pv (loose-leaf).*

Ajiboye, O.O., Yakubu, A.F. and Adams, T.E. 2012. A perspective on the ingestion and nutritional effects of feed additives in farmed fish species. World J. Fish Mar. Sci., 4, 87-101.

Anand, P.S., Kohli, M.P.S., Kumar, S., Sundaray, J.K., Roy, S.D., Venkateshwarlu, G., Sinha, A. and Pailan, G.H. 2014. Effect of dietary supplementation of biofloc on growth performance and digestive enzyme activities in *Penaeus monodon.* Aquaculture, 418: 108-115.

Anand, P.S., Kohli, M.P.S., Roy, S.D., Sundaray, J.K., Kumar, S., Sinha, A., Pailan, G.H. and kumar Sukham, M. 2013. Effect of dietary supplementation of periphyton on growth performance and digestive enzyme activities in *Penaeus monodon*. Aquaculture, 392: 59-68.

APHA, A. 2005. Standard methods for the estimation of water and waste water, 20th ed. Am. Public Health Assoc. Am. Water Works Assoc. Water Environ. Fed, Washington, D.C 21: 258-259.

Avnimelech, Y. and Kochba, M. 2009. Evaluation of nitrogen uptake and excretion by tilapia in bio floc tanks, using 15N tracing. Aquaculture, 287: 163-168.

Avnimelech, Y. and Ritvo, G. 2003. Shrimp and fish pond soils: processes and management. Aquaculture, 220: 549-567.

Avnimelech, Y. 1999. Carbon/nitrogen ratio as a control element in aquaculture systems. Aquaculture, 176: 227-235.

Avnimelech, Y. 2007. Feeding with microbial flocs by tilapia in minimal discharge bio-flocs technology ponds. Aquaculture, 264: 140-147.

Avnimelech, Y., Kochva, M. and Diab, S. 1994. Development of controlled intensive aquaculture systems with a limited water exchange and adjusted carbon to nitrogen ratio. Isr. J. Aquacult. Bamidgeh, 46: 119-131.

Ballester, E.L.C., Abreu, P.C., Cavalli, R.O., Emerenciano, M., De Abreu, L. and Wasielesky Jr, W. 2010. Effect of practical diets with different protein levels on the performance of *Farfantepenaeus paulensis* juveniles nursed in a zero exchange suspended microbial flocs intensive system. Aquacult. Nutr., 16(2): pp.163-172.

Boyd, C.E., Thunjai, T. and Boonyaratpalin, M. 2002. Dissolved salts in water for inland low-salinity shrimp culture. Global Aquacult. Advocate, 5: 40-45.

Bradford, M.M. 1976. A rapid and sensitive method for the quantitation of microgram quantities of protein utilizing the principle of protein-dye binding. Anal. Biochem., 72(1-2): 248-254.

Burford, M. A., Thompson, P. J., Mc Intosh, Bauman, R. H. and Pearson, D. C. 2003. Nutrient and microbial dynamics in high intensity, zero exchange shrimp ponds in Belize. Aquaculture, 219: 393 – 411.

Cherry, I.S. and Crandall Jr, L.A. 1932. The specificity of pancreatic lipase: its appearance in the blood after pancreatic injury. Am. J. Physiol. -Legacy Content, 100: 266-273.

Clark, J. M. 1964. Experimental Biochemistry, W. H. Freeman and Company, San Francisco and London, 228.

Crab, R. 2010. Bioflocs technology: an integrated system for the removal of nutrients and simultaneous production of feed in aquaculture (Doctoral dissertation, Ghent University).

Crab, R., Avnimelech, Y., Defoirdt, T., Bossier, P. and Verstraete, W. 2007. Nitrogen removal techniques in aquaculture for a sustainable production. Aquaculture, 270: 1-14.

De Schryver, P., Crab, R., Defoirdt, T., Boon, N. and Verstraete, W. 2008. The basics of bio-flocs technology: the added value for aquaculture. Aquaculture, 277: 125-137.

Decamp, O., Conquest, L., Forster, I. and Tacon, A.G.J. 2002. The nutrition and feeding of marine shrimp within zero-exchange aquaculture production system: role of Eukaryotic microorganisms. Microbial approaches to the aquatic nutrition within environmentally sound aquaculture production systems. World Aquaculture Society. Baton Rouge. USA, 86p..

Felix, S., Cheryl Antony, Gopalakannan. A., Antony Jesu Prabhu, P., Rajaram R. 2015. Intensive nursery rearing of *Litopenaeus vannamei* in Aerobic Microbial Floc (BIOFLOC) driven low saline raceway system. World Aquaculture 2015 - Meeting Abstract.

Funge-Smith, S.J. and Briggs, M.R. 1998. Nutrient budgets in intensive shrimp ponds: implications for sustainability. Aquaculture, 164: 117-133.

Gamboa delgado, J., Molina poveda, C. and Cahu, C. 2003. Digestive enzyme activity and food ingesta in juvenile shrimp *Litopenaeus vannamei* (Boone, 1931) as a function of body weight. Aquacult. Res., 34: 1403-1411.

Harris, J.M. 1993. The presence, nature, and role of gut microflora in aquatic invertebrates: a synthesis. Microb. Ecol., 25: 195-231.

Jackson, C., Preston, N., Thompson, P.J. and Burford, M. 2003. Nitrogen budget and effluent nitrogen components at an intensive shrimp farm. Aquaculture, 218: 397-411.

Ju, Z.Y., Forster, I., Conquest, L. and Dominy, W. 2008. Enhanced growth effects on shrimp (Litopenaeus vannamei) from inclusion of whole shrimp floc or floc fractions to a formulated diet. Aquacult. Nut., 14: 533-543.

Kuhn, D.D., Boardman, G.D., Lawrence, A.L., Marsh, L. and Flick Jr, G.J. 2009. Microbial floc meal as a replacement ingredient for fish meal and soybean protein in shrimp feed. Aquaculture, 296: 51-57.

Kuhn, D.D., Lawrence, A.L., Boardman, G.D., Patnaik, S., Marsh, L. and Flick Jr, G.J. 2010. Evaluation of two types of bioflocs derived from biological treatment of fish effluent as feed ingredients for Pacific white shrimp, *Litopenaeus vannamei*. Aquaculture, 303: 28-33.

Lara-Flores, M., Olvera-Novoa, M.A., Guzmán-Méndez, B.E. and López-Madrid, W. 2003. Use of the bacteria Streptococcus faecium and Lactobacillus acidophilus, and the yeast Saccharomyces cerevisiae as growth promoters in Nile tilapia (*Oreochromis niloticus*). Aquaculture, 216: 193-201.

Lee, P.G., Smith, L.L. and Lawrence, A.L. 1984. Digestive proteases of *Penaeus vannamei* Boone: relationship between enzyme activity, size and diet. Aquaculture, 42: 225-239.

Liu, L., Hu, Z., Dai, X. and Avnimelech, Y. 2014. Effects of addition of maize starch on the yield, water quality and formation of bioflocs in an integrated shrimp culture system. Aquaculture, 418: 79-86.

Liu, W.C., Luo, G.Z., Li, L., Wang, X.Y., Wang, J., Ma, N.N., Sun, D.C. and Tan, H.X. 2017. Nitrogen Dynamics and Biofloc Composition using Biofloc Technology to Treat Aquaculture Solid Waste Mixed with Distillery Spent Wash. N. Am. J. Aquac., 79: 27-35.

Lu, L., Tan, H., Luo, G. and Liang, W. 2012. The effects of Bacillus subtilis on nitrogen recycling from aquaculture solid waste using heterotrophic nitrogen assimilation in sequencing batch reactors. Bioresour. Technol., 124: 180-185.

Luo, G.Z., Avnimelech, Y., Pan, Y.F. and Tan, H.X. 2013. Inorganic nitrogen dynamics in sequencing batch reactors using biofloc technology to treat aquaculture sludge. Aquacult. Eng., 52, 73-79.

McGraw, W., Teichert-Coddington, D.R., Rouse, D.B. and Boyd, C.E. 2001. Higher minimum dissolved oxygen concentrations increase penaeid shrimp yields in earthen ponds. Aquaculture, 199: 311-321.

McGraw, W.J. and Scarpa, J. 2003. Minimum environmental potassium for survival of Pacific white shrimp *Litopenaeus vannamei* (Boone) in freshwater. J. Shellfish Res., 22: 263-268.

McIntosh, R.P. 2000. Changing paradigms in shrimp farming IV, low protein feeds and feeding strategies, Global Aquacult. Advocate, 3: 44-50.

Moss, S.M. 1994. Growth rates, nucleic acid concentrations, and RNADNA ratios of juvenile white shrimp, *Penaeus vannamei*, fed different algal diets. J. Exp. Mar. Biol. Ecol., 182: 193-204.

Moss, S.M., Arce, S.M., Argue, B.J., Otoshi, C.A., Calderon, F.R. and Tacon, A.G. 2001. Greening of the blue revolution: efforts toward environmentally responsible shrimp culture. In The New Wave, Proceedings of the special session on sustainable shrimp culture, Aquaculture, 1-19.

Moss, S.M., LeaMaster, B.R. and Sweeney, J.N. 2000. Relative Abundance and Species Composition of Gram Negative, Aerobic Bacteria Associated with the Gut of Juvenile White Shrimp *Litopenaeus vannamei* Reared in Oligotrophic Well Water and Eutrophic Pond Water. J. World Aquacult. Soc., 31: 255-263.

Raj, A., Menaga, M. and Felix, S. 2017. Effect of C/N ratio on water quality using distillery spentwash as a carbon source in indoor aerobic microbial floc tanks. Int. J. Sci. Environ. Technol., 6: 1865-1869.

Sahu, B.C., Adhikari, S. and Dey, L. 2013a. Carbon, nitrogen and phosphorus budget in shrimp (*Penaeus monodon*) culture ponds in eastern India. Aquacult. Int., 21: 453-466.

Sahu, B.C., Adhikari, S., Mahapatra, A.S. and Dey, L. 2013b. Carbon, nitrogen, and phosphorus budget in scampi (*Macrobrachium rosenbergii*) culture ponds. Environ. Monit. Assess., 185: 10157-10166.

Samocha, T.M., Patnaik, S., Speed, M., Ali, A.M., Burger, J.M., Almeida, R.V., Ayub, Z., Harisanto, M., Horowitz, A. and Brock, D.L. 2007. Use of molasses as carbon source in limited discharge nursery and grow-out systems for *Litopenaeus vannamei*. Aquac. Eng., 36: 184-191.

Sarath, G., De La Motte, R.S. and Wagner, F.W. 1989. Proteolytic enzymes: a practical approach. Oxford UK, 25.

Schneider, W.C. 1957. [99] Determination of nucleic acids in tissues by pentose analysis.

Schveitzer, R., Arantes, R., Costódio, P.F.S., do Espírito Santo, C.M., Arana, L.V., Seiffert, W.Q. and Andreatta, E.R. 2013. Effect of different biofloc levels on microbial activity, water quality and performance of *Litopenaeus vannamei* in a tank system operated with no water exchange. Aquac. Eng., 56: 59-70.

Serra, F.P., Gaona, C.A., Furtado, P.S., Poersch, L.H. and Wasielesky, W. 2015. Use of different carbon sources for the biofloc system adopted during the nursery and grow-out culture of *Litopenaeus vannamei*. Aquacult. Int., 23, 1325-1339.

Silva, E., Silva, J., Ferreira, F., Soares, M., Soares, R. and Peixoto, S. 2015. Influence of stocking density on the zootechnical performance of *Litopenaeus vannamei* during the nursery phase in a biofloc system. Boletim do Instituto de Pesca, 41: 777-783.

Sun, W. and Boyd, C.E. 2013. Phosphorus and nitrogen budgets for inland, saline water shrimp ponds in Alabama. Fish. Aquac. J., 4: 1–5.

Tewary, A. and Patra, B.C. 2008. Use of vitamin C as an immunostimulant. Effect on growth, nutritional quality, and immune response of *Labeo rohita* (Ham.). Fish Physiol Biochem., 34: 251-259.

UNDESA. 2014. United Nations Department of Economic and Social Affairs (2015): World Urbanization Prospects.

Van Wyk P and Scarpa, J. 1999. Water quality requirements and management in: Van Wyck P (ed) Farming marine shrimp in recirculating freshwater systems. Florida Department of Agriculture and Consumer Services, Tallahasee, 128–138.

Wilder, I.B. and Stanley, J.G. 1983. RNA DNA ratio as an index to growth in salmonid fishes in the laboratory and in streams contaminated by carbaryl. J. Fish Biol., 22: 165-172.

Xu, W.J. and Pan, L.Q. 2012. Effects of bioflocs on growth performance, digestive enzyme activity and body composition of juvenile *Litopenaeus vannamei* in zero-water exchange tanks manipulating C/N ratio in feed. Aquaculture, 356: 147-152.

Xu, W.J. and Pan, L.Q. 2013. Enhancement of immune response and antioxidant status of *Litopenaeus vannamei* juvenile in biofloc-based culture tanks manipulating high C/N ratio of feed input. Aquaculture, 412: 117-124.

Xu, W.J., Pan, L.Q., Zhao, D.H. and Huang, J. 2012. Preliminary investigation into the contribution of bioflocs on protein nutrition of *Litopenaeus vannamei* fed with different dietary protein levels in zero-water exchange culture tanks. Aquaculture, 350: 147-153.

Yao, C., Tan, H.X., Luo, G.Z. and Li, L. 2013. Effects of temperature on inorganic nitrogen dynamics in sequencing batch reactors using biofloc technology to treat aquaculture sludge. N. Am. J. Aquac., 75: 463-467.

Zhang, J., Liu, Y., Tian, L., Yang, H., Liang, G. and Xu, D. 2012. Effects of dietary mannan oligosaccharide on growth performance, gut morphology and stress tolerance of juvenile Pacific white shrimp, *Litopenaeus vannamei*. Fish Shellfish Immunol., 33: 1027-1032.

Ziaei-Nejad, S., Rezaei, M.H., Takami, G.A., Lovett, D.L., Mirvaghefi, A.R. and Shakouri, M. 2006. The effect of Bacillus spp. bacteria used as probiotics on digestive enzyme activity, survival and growth in the Indian white shrimp *Fenneropenaeus indicus*. Aquaculture, 252: 516-524.

CHAPTER 5

IN-VIVO STUDY OF *BACILLUS SP* ISOLATED FROM BIOFLOC SYSTEM IN GIFT TILAPIA

Biofloc technology is emerging as one of the sustainable technologies to increase the aquaculture production. It is known to possess several immunostimulatory compounds exhibiting possible probiotic effect in the culture of aquatic animals. The present study aimed to evaluate the *in vivo* efficiency of *Bacillus infantis* (T1), *Bacillus subtilis* (T2), *Exiguobacterium profundum* (T3) and *Bacillus megaterium* (T4) isolated from biofloc systems for improving the growth and immune performance of GIFT Tilapia. Animals (10±0.08g) were stocked at a density of 100 per m^{-3} in 500 liters FRP tanks for 42 days in triplicates. All the four probiotics (OD =1) were mixed with basal diet in treatments and feed without probiotic maintained as Control (C). A significant difference ($P< 0.05$) in weight gain, specific growth rate and FCR were observed between treatments and control with 100% survival. Serum albumin, globulin, protein, total blood count, glucose, myeloperoxidase activity and SOD were significantly different ($P < 0.05$) between treatments and control. T4 and T2 showed better immunological and anti-oxidant ability when compared to other strains. Results from principal component analysis demonstrated that *B. megaterium* and *B. subtilis* can be the promising probiotic bacteria isolated from biofloc systems exhibiting multiple benefits with improved growth and health of the culture animals.

INTRODUCTION

Tilapias, the second most farmed freshwater fish, have intrinsic feature like fast growth rate, disease resistance ability, low trophic level feeding and good flesh quality. The total production of tilapia in was 6.8 million tonnes in 2018 (FAO, 2020) and expected to produce 7.3 million tonnes by 2030 (Behera *et al.*, 2018). The rapid expansion and intensification of aquaculture resulted in the outbreak of diseases leading to considerable economic losses and thereby hindering the sustainable development of the industry (Rico *et al.*, 2014 and Barria *et al.*, 2020).

To prevent and treat diseases in aquatic animals, antibiotics have been used for improving aquaculture production. Indiscriminate-usage of antibiotics may result in the development of antibiotic resistant bacteria, antibiotic residues in the flesh and the microbial population destruction in the aquatic environment (Marques *et al.*, 2015 and Kuebutornye *et al.*, 2020). This leads the adoption of various alternative strategies to replace the use of antibiotics with pre and probiotics.

As probiotics gained its success in human and animal feeding practices now the attention is on aquaculture. Probiotics, live microbes when administered as a dietary supplement in *Oreochromis niloticus*, confer benefits to the host by improving the balance of the intestinal microbiota, exclusion of pathogens and growth performance. The commercially available probiotics used for tilapia culture were isolated from a wide range of sources such as tilapia culture water, sediment and its intestine (Apún Molina *et al.*, 2009; Aly *et al.*, 2008; El-Rhman *et al.*, 2009 and Del'Duca *et al.*, 2013). *Bacillus sp.* is one of the most commonly used probiotics compared to other species due to its active role in enhancing immune mechanism of tilapia (He *et al.*, 2013). *Bacillus* sp. has the efficient enzymatic pathway in breaking the complex carbohydrates, proteins and lipids (Menaga *et al.*, 2017) and its spore- forming ability (Hong *et al.*, 2005) adds an extra advantage of using it in the commercial aquaculture practices. In recent years, adoption of biofloc technology, a minimal water exchange technology promotes heterotrophic bacteria has replaced the commercial probiotics usage as it exerts the possible probiotic effect in various aquaculture animals (Hargreaves, 2013). However, reports on the identification and isolation of *Bacillus* sp. from biofloc based tilapia culture are limited. This study attempts to determine the efficiency of 4 *Bacillus* sp. namely *Bacillus subtilis, B. megaterium, B. infantis* and *Exiguobacterium profundum* from biofloc systems of GIFT tilapia. This study also involves the strict process of selection of probiotics from biofloc systems using *in vivo* test associated with extensive evaluation in different aspects such as immunological, haematological parameters and antioxidant status of GIFT tilapia.

MATERIALS AND METHODS

Isolation of Probiotic Bacteria from Biofloc Culture Water

The biofloc was developed and maintained in the 500 litres FRP tank according to Taw, 2006. Distillery spentwash as a carbon source was used to maintain C: N ratio at 10:1. The biofloc water sample was fortnightly screened for probiotic bacteria by spread plating on MRS agar. The morphologically different isolates were characterized according to Bergey's Manual of Determinative Bacteriology. DNA was isolated from the bacterial culture using phenol- chloroform method later amplified for its 16s rRNA region using Forward primer - 5'AGAGTTTGATCCTGG CTCAG3' and Reverse primer- 5'CGGTTACCTTGTTACG ACTT3'. The amplified DNA was sequenced using Sanger's method and obtained sequences were aligned and submitted in Genbank. The accession numbers for the *Bacillus* sp. were obtained as *Bacillus infantis*- MH424755 *Bacillus subtilis*- MH424900, *Exiguobacterium profundum*- MH424898 and *Bacillus megaterium*- MH424904.

Experimental Design

A commercial fish feed (30% protein, 4% fat, 4% fibre and 14% ash of Grobest Feed, India) was used as the basal diet. The experimental diet includes the basal diet supplemented with four bacterial cultures isolated from biofloc water which were grown in LB medium (37 °C for 16 h). The cell pellets were washed and resuspended in PBS to attain the OD value of 1.00. Later bacterial suspension was homogenized and sprayed on the experimental diets at a rate of 100ml bacterial suspension/kg of feed (Zokaeifar *et al.*, 2012). Without bacteria the same volume of PBS was added to the basal diet for control. The viability of bacteria in the experimental diets remains stable for seven days based on the confirmation with plating on the nutrient agar. Hence, these diets were prepared weekly once to warrant the bacterial performance.

GIFT tilapia juveniles (10+0.08g) were stocked in fifteen FRP tanks of 500 litres capacity (50 animals per tank) filled with freshwater. The experiment was conducted for 42 days and treatments include: Control (basal diet without bacteria), T-1 (basal diet supplemented with *Bacillus infantis*), T-2 (basal diet supplemented with *Bacillus subtilis*), T-3 (basal diet supplemented with *Exiguobacterium profundum*) and T-4 (basal diet supplemented with *Bacillus megaterium*) in triplicates. The animals were fed with two rations at 3% of average body weight throughout the experimental trial. Regular siphoning of water from the experimental units was carried out to maintain the optimum water quality.

Growth Performance

The growth performance and survival of juvenile fishes for all groups were calculated using the following equations:

$$\text{Weight gain (WG in g)} = W_t - W_0$$

$$\text{Specific growth rate (SGR, \% day}^{-1}) = [\ln W_t - \ln W_0] / t \times 100$$

FCR = feed offered (dried weight)/weight gain (wet weight);

$$\text{Survival rate (in \%)} = (Nt \times 100) / N_0$$

Where W_0 and W_t are the initial weight (g) and final weight (g) of fish, N_0 and N_t are initial and final number of fish, and t represents culture duration in days.

Water Quality Parameters

Temperature (mercury thermometer) and pH (Labtronics pH meter) were measured on a daily basis. As per APHA (2008), parameters like dissolved oxygen, free carbon dioxide, alkalinity, hardness, calcium and magnesium ion concentration were measured weekly. Nitrate-N (NO_3 -N) and nitrite nitrogen (NO_2 -N) were estimated using the filtered water samples and analysed using Resorcinol method. Ammonia was estimated by phenol hypochlorite method and orthophosphate according to APHA (2008) every week.

Haematological, Immunological and Anti-Oxidant Indicators

Fishes were anesthetized using clove oil and the blood samples were drawn from the caudal vein. The blood was then transferred to EDTA (an anticoagulant) coated vials and for serum separation the blood sample was allowed to clot for 15min, centrifuged and used for further analysis.

Total Blood Count, Albumin and Globulin Content

The total blood count of the experimental animals was enumerated using Giemsa staining method.

To the 10µL of serum, 2.5mL of reagent R containing 200mM/L of succinate buffer, 0.4mM/L of bromocresol green and 4mM/L of sodium azide was added. 10µL of albumin (4g/dL) served as standard. The mixture was incubated for 5min at 20-25°C. The absorbance was read at 623nm against blank (Doumas *et al.*, 1971).

$$\text{Albumin concentration (g/dL)} = (A_{specimen} / A_{standard}) \times 4$$

$$\text{Globulin} = \text{Total serum protein} - \text{albumin}$$

Respiratory Burst Activity

Respiratory burst activity was performed following the modified method of Anderson and Siwiki (1995). 0.2% of nitroblue tetrazolium (NBT) solution was added to the 0.1mL blood sample and incubated for 30min at room temperature (30°C). 1.0 mL N, N-dimethyl formamide (DMF) was added to the 0.05mL of the NBT blood cell suspension and centrifuged for 5 min at 5000 rpm. The collected supernatant was read on a spectrophotometer at 540 nm.

Myeloperoxidase Activity

Total MPO content present in serum was measured according to Quade and Roth (1997) with slight modification. Ten µL of serum sample collected was diluted with 90 µL of Hank's balanced salt solution (HBSS). 35 µL of 20 mM 3,3',5,5'-tetramethyl benzidine hydrochloride (TMB) and five mM H_2O_2 (freshly prepared) were added to the serum. The colour change reaction was stopped after 2 min by adding 35 µL of 4 M sulphuric acid (H_2SO_4) and the OD was read at 450 nm.

Units/mL enzyme = (A470nm Test at 1 min - A470nm Blank at 1 min) (df) / (1.0) (0.035)

df = Dilution factor

1.0 = The increase in A470nm/minute per unit of enzyme

0.035 = Volume of enzyme used (mL)

Glucose and Protein Estimation

The serum sample was analysed for glucose level using Beacon diagnostics pvt. Ltd., kit. The protein estimation in blood as well in serum was done by Lowry's method (Lowry *et al.*, 1951).

Catalase stress Enzyme Assay

The blood sample (10-50μL) was added to 2.5mL of phosphate buffer (50mM/ pH7). 1mL of 0.3% H_20_2 (freshly prepared) was added to the above suspension. The decrease in the absorbance was read at 240 nm for 3mins at 30 sec interval (Takahara *et al.*, 196).

CAT (units/mg protein) = [OD/min (3) x total volume]/ [34 x sample volume x protein] x 1000

Super Oxide Dismutase (SOD) Assay

The blood sample (10-50μl) was added to 1.5mL of carbonate buffer (0.1 M/ pH10.2). 0.5mL of epinephrine (freshly prepared) was added to the above suspension. The increase in the absorbance was read at 480 nm for 3mins at 30 sec interval

(Misra *et al.*, 1972).

Inhibition % = OD blank-change in OD/min/OD blank x 100

SoD unit/mg of protein = inhibition %/50 x sample vol x 18 x mg of protein

One-way ANOVA was performed to find the significant difference between the treatments and control using SPSS version 20.0 for the immunological, haematological, growth and water quality parameters. Statistical difference was found at $P < 0.05$. The principal component analysis was performed to find the overall immune performance of probiotic isolates.

RESULTS AND CONCLUSIONS

Growth Performance

The weight gain, specific growth rate and feed conversion ratio of GIFT tilapia fed with different experimental diets along with the statistical analysis were shown in the table 1. All the growth parameters showed significant difference between control and treatments ($P<0.05$). *Bacillus megaterium* and *Bacillus subtilis* fed animals possessed higher weight gain with improved FCR.

Water Quality Parameters

The various water quality parameters along with the statistical analysis were shown in the table 2. There was no significant difference between control and treatments for all the parameters analysed.

Table 1. Growth Performance of GIFT Tilapia fed with different experimental diets at the end of the culture trial

Parameters	Control	*B. infantis*	*B. subtilis*	*E. profundum*	*B. megaterium*
Initial Weight(gm)	10 ± 0.05^{a}	10 ± 0.08^{a}	10 ± 0.06^{a}	10 ± 0.02^{a}	10 ± 0.04^{a}
Final Weight(gm)	20.6 ± 1.13^{a}	26.3 ± 1.64^{b}	27.2 ± 0.81^{b}	25.9 ± 0.32^{b}	27.96 ± 1.00^{b}
Weight gain (gm)	10.6 ± 1.13^{a}	16.3 ± 1.64^{b}	17.2 ± 0.81^{b}	15.9 ± 0.32^{b}	17.96 ± 1.00^{b}
Specific growth rate	1.73 ± 0.16^{a}	2.28 ± 0.33^{b}	2.41 ± 0.15^{b}	2.29 ± 0.11^{b}	2.45 ± 0.16^{b}
Feed conversion ratio	2.55 ± 0.15^{a}	2.12 ± 0.24^{b}	2.1 ± 0.2^{b}	2.21 ± 0.18^{b}	2.09 ± 0.24^{b}

Data assigned with different superscripts denote significant difference in a row ($P < 0.05$)

Table.2 - Water quality parameters of experimental groups in the 42 days culture trial of GIFT Tilapia

Parameters	Control	*B. infantis*	*B. subtilis*	*E. profundum*	*B. megaterium*
pH	7.24 ± 0.1^{a} (7.12 – 7.36)	7.22 ± 0.15^{a} (7.14 – 7.3)	7.37 ± 0.17^{a} (7.33 – 7.79)	7.33 ± 0.21^{a} (7.16 – 7.5)	7.42 ± 0.22^{a} (7.25 – 7.59)
Temperature (°C)	27.45 ± 0.19^{a} (27– 30)	27.26 ± 0.12^{a} (27 – 30)	27.45 ± 0.1^{a} (27– 30)	27.26 ± 0.12^{a} (27 – 30)	27.56 ± 0.17^{a} (27 – 30)
DO (mg/L)	5.45 ± 0.04^{a} (4.1- 5.7)	5.7 ± 0.04^{a} (4.2 – 5.9)	5.55 ± 0.04^{a} (4.3 - 5.7)	5.72 ± 0.04^{a} (4.5 - 5.9)	5.65 ± 0.04^{a} (4.4 - 5.8)
Free carbon di oxide (mg/L)	2.4± 0.15^{a} (1.9 - 3.2)	2.33 ± 0.1^{a} (1.7 - 3.3)	2.65 ± 0.15^{a} (1.7 - 3.4)	2.77 ± 0.15^{a} (1.9 – 3.5)	2.55 ± 0.12^{a} (1.8- 3.3)
Alkalinity (mg/L)	55 ± 2.5^{a} (45 – 65)	57.89 ± 2.9^{a} (38 – 79)	56.8 ± 2.45^{a} (36 – 65)	57.25 ± 2.6^{a} (39 – 78)	56.89 ± 2.75^{a} (37 – 75)
Hardness (mg/L)	250 ± 3.6^{a} (236 – 292)	249 ± 5.6^{a} (219 – 279)	252 ± 3.9^{a} (225 – 277)	248 ± 5.1^{a} (210 – 269)	251 ± 5.15^{a} (241 – 267)
Calcium ions (mg/L)	55.59 ± 4.75^{a} (44 – 75)	55 ± 3.77^{a} (34 – 65)	53.59 ± 4.54^{a} (46 – 69)	53.1 ± 4.98^{a} (34 – 68)	54.5 ± 3.98^{a} (38 – 70)
Magnesium ions (mg/L)	79 ± 2.7^{a} (45 – 85)	78 ± 2.7^{a} (42 – 88)	79.55 ± 2.74^{a} (42 – 80)	77.5 ± 2.55^{a} (41– 85)	78.12 ± 2.9^{a} (42 – 88)
Nitrate (mg/L)	0.007 ± 0.0^{a} (0.004 – 0.012)	0.006 ± 0.0^{a} (0.004 – 0.015)	0.007 ± 0.0^{a} (0.005 – 0.008)	0.006 ± 0.0^{a} (0.003 – 0.01)	0.008 ± 0.0^{a} (0.004 – 0.015)
Nitrite (mg/L)	0.014 ± 0.0^{a} (0.007 – 0.019)	0.011 ± 0.0^{a} (0.006 – 0.015)	0.013 ± 0.0^{a} (0.008 – 0.019)	0.012 ± 0.0^{a} (0.005 – 0.015)	0.011 ± 0.0 ^{a} (0.004 – 0.015)
Ammonia (mg/L)	0.07 ± 0.02^{a} (0.05 – 0.19)	0.05 ± 0.025^{a} (0.04– 0.26)	0.04 ± 0.03^{a} (0.04 – 0.19)	0.06 ± 0.032^{a} (0.04– 0.29)	0.05 ± 0.024^{a} (0.05– 0.25)
Phosphate (mg/L)	0.08 ± 0.02^{a} (0.05 – 0.15)	0.085 ± 0.01^{a} (0.04 – 0.17)	0.082± 0.02^{a} (0.033 – 0.12)	0.09 ± 0.015^{a} (0.044 – 0.16)	0.083 ± 0.018^{a} (0.04 – 0.19)

Data assigned with different superscripts denote significant difference in a row ($P < 0.05$)

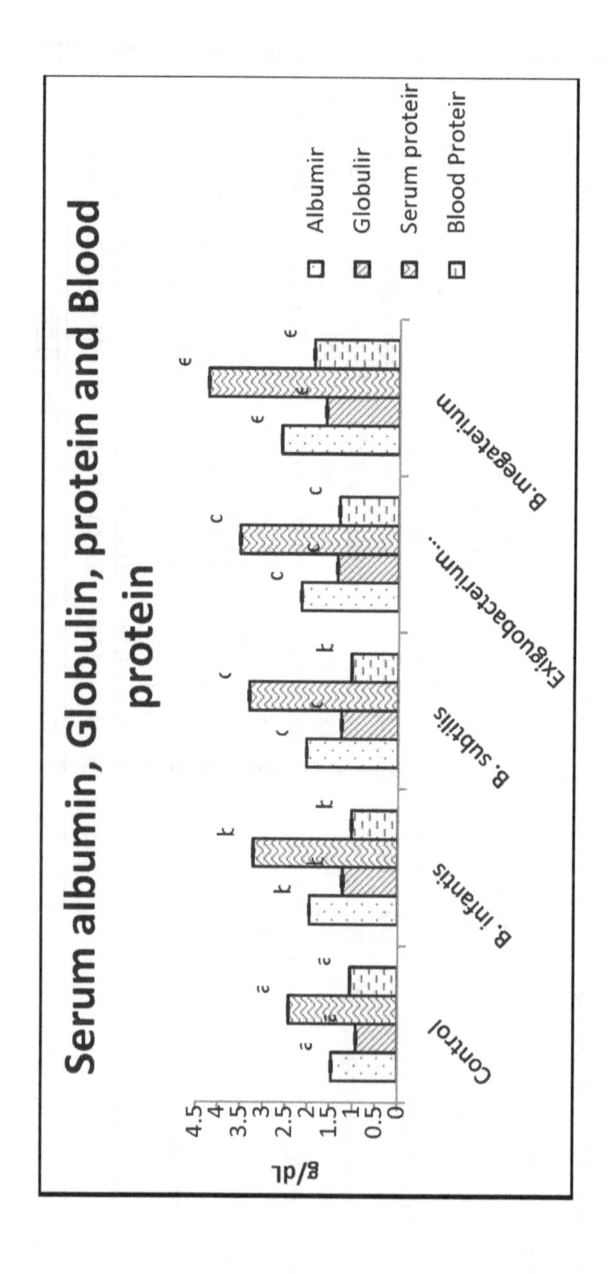
Serum albumin, Globulin, protein and Blood protein
g/dL
0
0.5
1
1.5
2
2.5
3
3.5
4
4.5
Control
B. infantis
B. subtilis
Exiguobacterium...
B.megaterium
Albumir
Globulir
Serum proteir
Blood Proteir

Haematological, Immunological and Anti-Oxidant Indicators

The immunological and haematological parameters were analysed and the graphs along with the standard deviation were constructed which were represented in the figure 1.

From the One way ANOVA performed, the results confer that there is a significant difference between treatments (T1, T2, T3 and T4) and control (C) for serum albumin, globulin, protein, total blood count, glucose, myeloperoxidase and SOD. No significant difference was found in the blood protein levels of animals fed with *Bacillus infantis* and *Bacillus subtilis* diets. Respiratory burst activity of the experimental animals fed with *Bacillus subtilis* and *Exiguobacterium profundum* showed no significant differences.

The first four principal components (PCs) accounted for 92.45% of the total variance which was listed in the table 3. The distribution plot variables of PC 1 and PC 2 on the plane, which are mainly related to serum protein and myeloperoxidase are presented in the figure.2. These two selected components made up to 50.79 and 17.40% of the variance in the model, suggesting that they were key factors in distinguishing the different strains used. It could be inferred that strains presented in quadrant IV were significant as they showed high correlation with respect to variables. *Bacillus megaterium* and *B. subtilis* had the highest contribution to PC 1 and may be preferentially used as potential probiotics among all the strains.

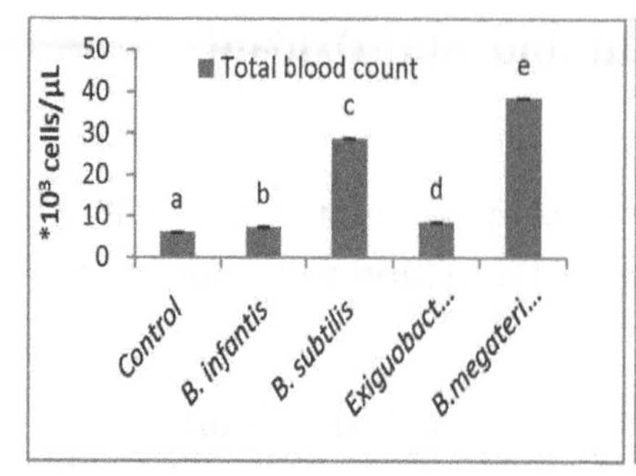

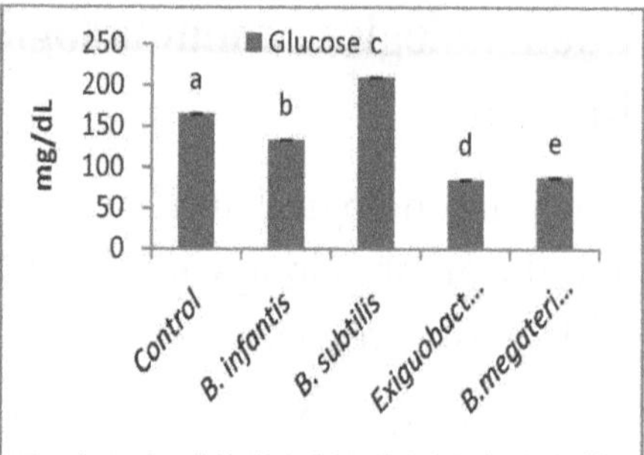

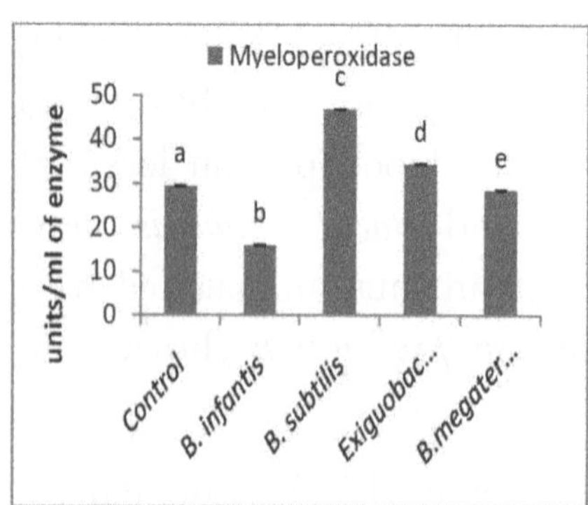

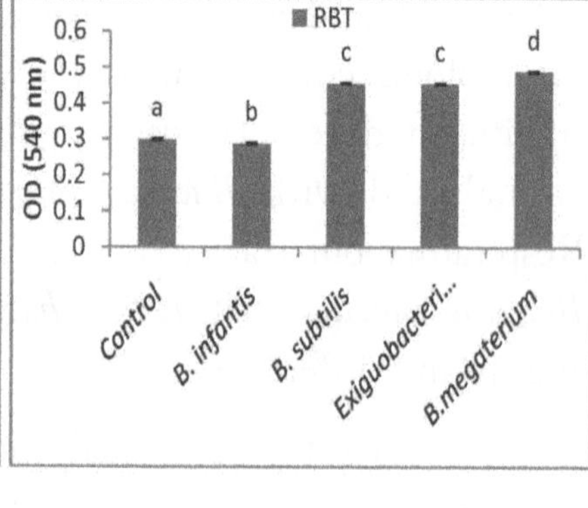

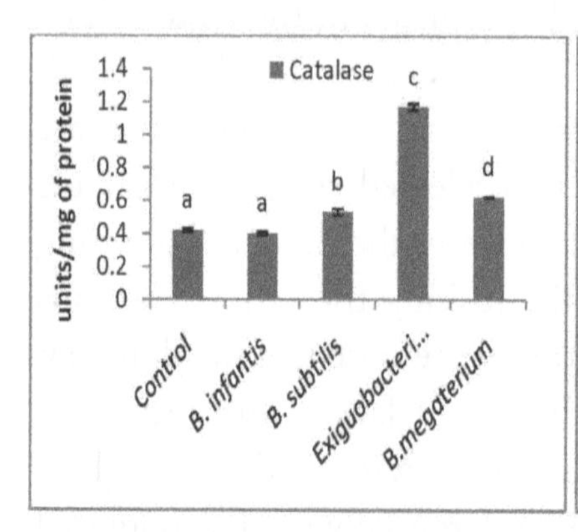

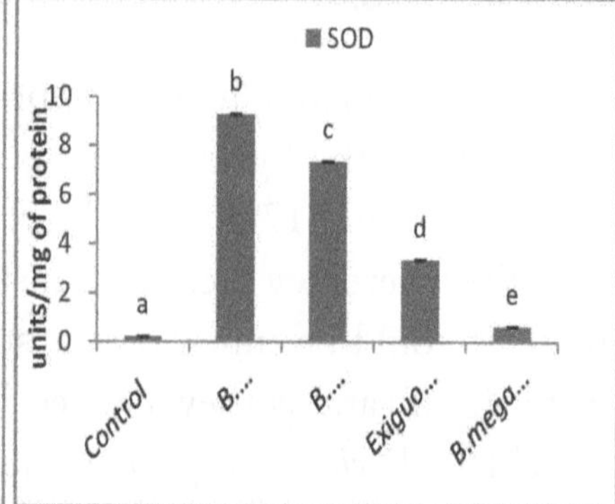

Fig.1. Haematological, Immunological and Anti Oxidant Indicators of GIFT Tilapia in the *in vivo* feeding trial of 42 days.

Data assigned with different superscripts denote the significant difference ($P < 0.05$).

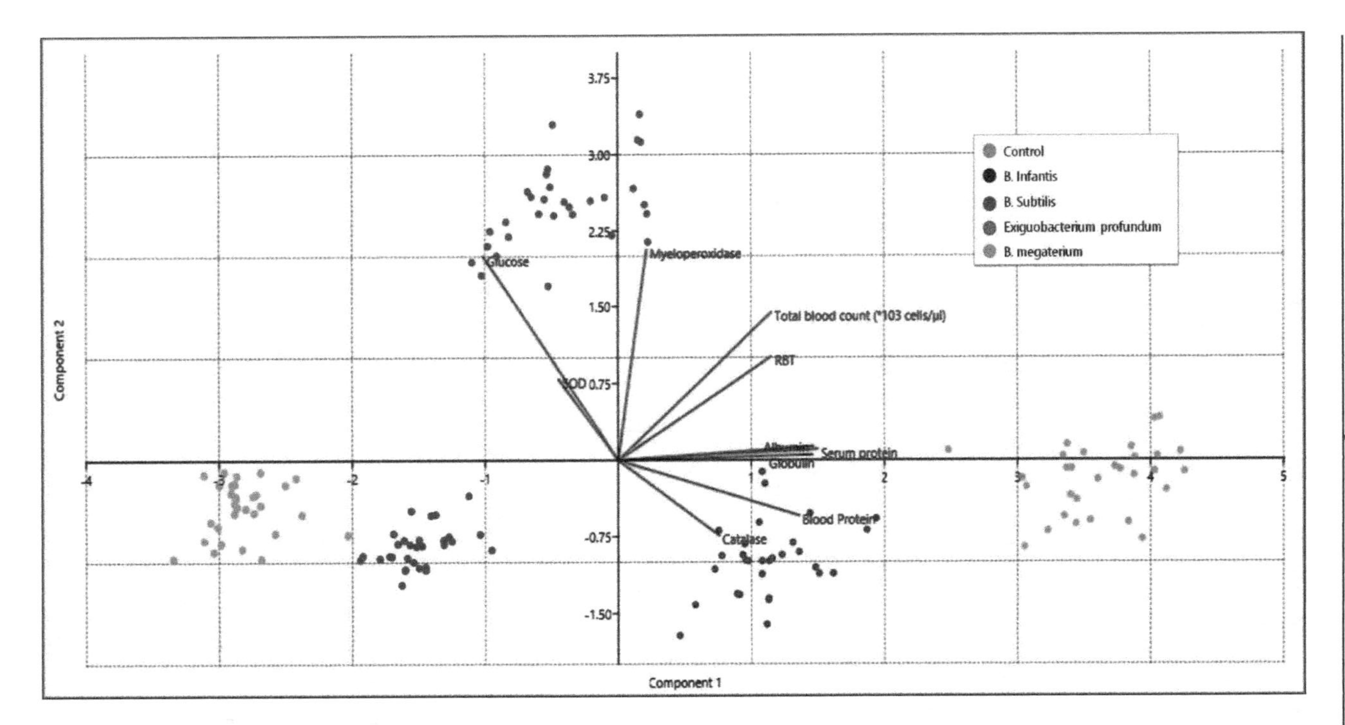

Fig. 2. Principal component analysis based on four probiotic strains and tested parameters

Table 3: Correlation of variables obtained from total variance with the factors of the PCA analysis

	PC 1	PC 2	PC 3	PC 4
Albumin	0.408	0.037	0.201	0.129
Globulin	0.406	0.014	0.195	0.133
Serum protein	0.417	0.030	0.200	0.114
Blood Protein	0.379	-0.152	0.016	-0.373
Total blood count (*103 cells/μl)	0.321	0.402	0.193	-0.312
Glucose	-0.283	0.558	0.071	-0.098
Myeloperoxidase	0.060	0.575	-0.476	0.069
RBT	0.321	0.281	-0.275	0.105
Catalase	0.211	-0.206	-0.541	0.498
SOD	-0.125	0.221	0.495	0.665
% variance	50.79	17.40	13.22	11.04

Growth performance of GIFT tilapia fed with probiotics showed positive results and ascertained with the fact of early findings (Avella *et al.*, 2010; Hoseinifar *et al.*, 2015; Doan *et al.*, 2018; Liu *et al.*, 2012 and Liu *et al.*, 2017). Higher growth rate was found in *B. megaterium* fed animals and lower growth rate was seen in control. 100% survival in all the experimental groups was observed and this may be due to the regular water exchange favouring good environmental condition for the animals. Water quality parameters showed no significant differences with the supplementation of probiotics along with the feed.

The haematological parameters such as albumin, globulin and total blood count were found to be more pronounced in probiotics fed tilapia than control. Albumin and globulins are the two major groups of serum proteins and taken as total protein. Blood globulin contents are inevitable for the healthy culture of animals with improved immune functions. However, for sustaining the osmotic pressure, albumin is the vital need for proper distribution of body fluids as it acts as a plasma carrier (Jha *et al.*, 2007). High concentrations of total protein in fish serum can be correlated with enhancement of the non-specific immune response. The results indicated higher ($P < 0.05$) albumin and globulin concentrations of tilapia blood serum were obtained in probiotic fed fishes compared with that of the control in the present study. Higher levels of albumin and globulin were found in *Bacillus megaterium* and this may be due to the presence of carotenoids and its antioxidant potential (Mitchell *et al.*, 1986). The increase in blood cell count in tilapia due to probiotic feed indicates an immunostimulant effect. The possible explanation for the variation of blood cells could be attributed to the various probiotic feed.

Immunological parameters such as respiratory burst and myeloperoxidase activity was significantly higher in the probiotic mixed diets than control. Immune systems can be activated in several ways like enhancing the number of phagocytes, activating phagocytes or increasing the synthesis of the involved molecules. Increased bacterial pathogen killing ability of phagocytes can be inferred from increased respiratory burst activity which is a most important bactericidal mechanism in fishes (Sharp *et al.*, 1993). It had been demonstrated that the RBT of tilapia when administered with bacterial probiotics isolated from biofloc culture systems showed an improved performance than control. Similar results were observed in the administration of *Lactobacillus rhamnosus* (strain ATCC 53103) in rainbow trout (*Oncorhynchus mykiss*) (Nikoskelainen *et al.*, 2003). The myeloperoxidase (MPO) utilizes one of the oxidative radicals to produce hypochlorous acid thereby acting as an antimicrobial enzyme. The MPO was mostly released by the azurophilic granules of neutrophils during oxidative respiratory burst. The improved MPO activity was seen in treatment with addition of *B. subtilis* as a probiotic to tilapia. This was in concurrent with the findings of Wang *et al.* (2008) on the addition of *E. faecium* ZJ4 to the aquaria.

Anti-oxidant status such as glucose, SOD and catalase activity was significantly different in the treatment than control. Intensive glycogenolysis and the synthesis of glucose from extra hepatic tissue proteins and amino acids increases the glucose content in blood as an indicator of stress in animals (Almeida *et al.*, 2001; Nakno and Tomlinson *et al.*, 1967) observed that all types of stress elevated the secretion of catecholamine which stimulates the breakdown of glycogen thereby increasing blood glucose level. In the present study, *B. megaterium* fed tilapia has less glucose level when compared with other treatments indicating the lesser stress levels in animals.

SOD and catalase are two important enzymes in the cellular antioxidant defence system, means of dealing with oxidative stress. SOD and catalase remove oxygen radicals produced within the cells (Kurata *et al.*, 1993) and are expected to increase under hypoxia to detoxify ROS. Lower levels of SOD and catalase are indication for cell damage due to the accumulation of the high-level free radical. The experimental animals fed with probiotic supplemented diets possessed increased SOD and catalase levels compared to control. However, within the treatments,

Exiguobacterium profundum and *Bacillus infantis* supplemented diets enhanced the SOD and catalase levels of the animals. The overall performance of the *Bacillus* sp. examined using Principal component analysis (PCA) of *in vivo* characteristics particularly immunological, haematological and anti-oxidant status revealed *Bacillus megaterium* and *Bacillus subtilis* as a superior strain with desirable qualities compared to the other strains for the GIFT tilapia culture. The results from the present study also confirmed the probiotic effect of biofloc as the strains were isolated from the biofloc systems probing the ecological health of culture animals.

The four bacterial species isolated from the biofloc water sample were tested for *in- vivo* probiotic efficiency in GIFT tilapia. Improved performance of the culture animals was found in probiotics fed animal and among the four bacillus strains, *Bacillus megaterium* and *B. subtilis* can be used as potential probiotics as they exhibited better performance in terms of immunological, haematological and growth parameters throughout the experimental trial. These two strains can further be commercialised and used as effective probiotics in aquaculture. The different physical forms of the tested probiotics and its combined performance would further help to determine its efficiency for the mass scale production in aquaculture.

ACKNOWLEDGEMENT

The authors express their heartfelt gratitude to Department of Biotechnology, Government of India, New Delhi for financially supporting this research project (DBT-Project Code: SU7; PI: Dr. S. Felix).

REFERENCES

Almeida, J.A., Novelli, E.L.B, Dal-Pai Silva, M. and Alves, Jr., R. 2001. Environmental cadmium exposure and metabolic responses of the Nile tilapia Oreochromis niloticus. *Environ. Pollut.* 114: 169–175.

Aly, S.M., Abd-El-Rahman, A.M., John, G. and Mohamed, M.F. 2008. Characterization of some bacteria isolated from Oreochromis niloticus and their potential use as probiotics. *Aquaculture*, 277: 1-6.

Anderson, D.P. and Siwicki, A.K. 1995. Basic hematology and serology for fish health programs.

APHA (American Public Health Association). 2008. Standard Methods for the Examination of Water and Wastewater. *American Public Health Association, Washington, DC.*

Apún Molina, J.P., Santamaría, Miranda, A., Luna González, A., Martínez Díaz, S.F. and Rojas Contreras, M. 2009. Effect of potential probiotic bacteria on growth and survival of tilapia Oreochromis niloticus L., cultured in the laboratory under high density and suboptimum temperature. *Aquaculture Research*, 40: 887-894.

Avella, M.A., Gioacchini, G., Decamp, O. 2010. Makridis P, Bracciatelli C & Carnevali O, Application of multi-species of *Bacillus* in sea bream larviculture. *Aquaculture*, 305: 12-19.

Barría, A., Trinh, T.Q., Mahmuddin, M., Benzie, J.A., Chadag, V.M. and Houston, R.D. 2020. Genetic parameters for resistance to Tilapia Lake Virus (TiLV) in Nile tilapia (Oreochromis niloticus). *Aquaculture*, 522: 735126.

Behera, B.K., Pradhan, P.K., Swaminathan, T.R., Sood, N., Paria, P., Das, A., Verma, D.K., Kumar, R., Yadav, M.K., Dev, A.K.,

Parida, P.K., Das, B.K., Lal, K.K. and Jena, J.K. 2018. Emergence of Tilapia Lake Virus associated with mortalities of farmed Nile Tilapia Oreochromis niloticus (Linnaeus 1758) in India. *Aquaculture*, 484: 168-174.

Del'Duca, A., Cesar, D.E., Diniz, C.G. and Abreu, P.C. 2013. Evaluation of the presence and efficiency of potential probiotic bacteria in the gut of tilapia (Oreochromis niloticus) using the fluorescent in situ hybridization technique. *Aquaculture*, 388: 115-121.

Doan, H.V., Hoseinifar, S.H., Khanongnuch, C., Kanpiengjai, A., Unban, K., Kim, V.V. and Srichaiyo, S. 2018. Host-associated probiotics boosted mucosal and serum immunity, disease resistance and growth performance of Nile tilapia (*Oreochromis niloticus*). Aquaculture, 491: 94-100.

Doumas, B.T., Watson, W.A. and Biggs, H.G. 1971. Albumin standards and the measurement of serum albumin with bromocresol green. *Clinica chimica acta*, 31: 87-96.

El-Rhman, A.M.A., Khattab, Y.A. and Shalaby, A.M. 2009. Micrococcus luteus and Pseudomonas species as probiotics for promoting the growth performance and health of Nile tilapia, Oreochromis niloticus. *Fish & Shellfish Immunology*, 27: 175-180.

FAO. 2020. The State of World Fisheries and Aquaculture 2020, Sustainability in action, Rome.

Hargreaves, J.A. 2013. Biofloc production systems for aquaculture. *SRAC*, 4503: 1-12.

He, S., Zhang, Y., Xu, L., Yang, Y., Marubashi, T., Zhou, Z. and Yao, B. 2013. Effects of dietary Bacillus subtilis C-3102 on the production, intestinal cytokine expression and autochthonous

bacteria of hybrid tilapia Oreochromis niloticus@&×
Oreochromis aureusB&. *Aquaculture*, 412: 125-130.

Hong, H.A., Duc, L.H. and Cutting, S.M. 2005. The use of bacterial spore formers as probiotics. *FEMS Microbiol Rev.* 29: 813-835.

Hoseinifar, S.H., Roosta, Z., Hajimoradloo, A. and Vakili, F. 2015. The effects of *Lactobacillus acidophilus* as feed supplement on skin mucosal immune parameters, intestinal microbiota, stress resistance and growth performance of black swordtail (*Xiphophorus helleri*). *Fish & shellfish immunology*, 42: 533-538.

Jha, A.K., Pal, A.K., Sahu, N.P., Kumar, S. and Mukherjee, S.C. 2007. Haemato-immunological responses to dietary yeast RNA, w-3fatty acid and b-carotene in Catla catla juveniles. *Fish & Shellfish Immunology*, 23: 917–927.

Kuebutornye, F.K. and Abarike, E.D. 2020. The contribution of medicinal plants to tilapia aquaculture: a review. *Aquaculture International*, 1-19.

Kurata, M., Suzuki, M. and Agar, N.S. 1993 Antioxidant systems and erythrocyte life-span in mammals, Comp. *Biochem. Physiol. B*, 106: 477–487.

Liu, H., Wang, S., Cai, Y., Guo, X., Cao, Z., Zhang, Y., Liu, S., Yuan, W., Zhu, W., Zheng, Y., Xie, Z., Guo, W. and Zhou, Y. 2017 Dietary administration of *Bacillus subtilis* HAINUP40 enhances growth, digestive enzyme activities, innate immune responses and disease resistance of tilapia, *Oreochromis niloticus. Fish Shellfish Immunol.*, 60: 326-333.

Liu,C.H., Chiu, C.H., Wang, S.W. and Cheng W. 2012. Dietary administration of the probiotic, *Bacillus subtilis* E20, enhances the growth, innate immune responses, and disease resistance of the grouper, *Epinephelu scoioides. Fish Shellfish Immunol.*, 33: 699-706.

Lowry, O.H., Rosebrough, N.J., Farr, A.L. and Randall, R.J. 1951. *J. Biol. Chem.*, 193: 265-275.

Marques, A., Dinh, T., Ioakeimidis, C., Huys, G., Swings, J., Verstraete, W., Dhont, J., Sorgeloos, P. and Bossier, P. 2005. Effects of bacteria on *Artemia franciscana* cultured in different gnotobiotic environments. *Applied and Environmental Microbiology*, 71: 4307-4317.

Menaga, M., Felix, S. and Gopalakannan, A. 2017. Identification of Bacterial Isolates from Aerobic Microbial Floc Systems by 16sr DNA Amplification and Sequence Analyses. *Indian Vet. J.*, 94: 24 – 26.

Misra, H.P. and Fridovich, I. 1972. The role of superoxide anion in the autoxidation of epinephrine and a simple assay for superoxide dismutase. *Journal of Biological chemistry*, 247: 3170-3175.

Mitchell, C., Iyer, S., Skomurski, J.F. and Vary, J.C. 1986. Red pigment in Bacillus megaterium spores. *Appl. Environ. Microbiol.*, 52: 64-67.

Nakno, T. and Tomlinson, N. 1967. Catecholamine and carbohydrate concentrations in rainbow trout (Salmo gairdneri) in relation to physical disturbance. *J. Fish. Res. Bd. Can.* 24: 1701–1715.

Nikoskelainen, S., Ouwehand, A., Bylund, G., Salminen, S. and Lilius, E.M. 2003. Immune enhancement in rainbow trout (Oncorhynchus mykiss) by potential probiotic bacteria (Lactobacillus rhamnosus). *Fish & Shellfish Immunology*, 15: 443-452.

Quade, M.J. and Roth, J.A. 1997. A rapid, direct assay to measure degranulation of bovine neutrophil primary granules. *Veterinary Immunology and Immunopathology*, 58: 239-248.

Rico, A., Oliveira, R., McDonough, S., Matser, A., Khatikarn, J., Satapornvanit, K., Nogueira, A.J., Soares, A.M., Domingues, I. and Brink, P.J. 2014. Use, fate and ecological risks of antibiotics applied in tilapia cage farming in Thailand. *Environmental Pollution*, 191: 8-16.

Sharp, G.J.E. and Secombes, C.J. 1993. The role of reactive oxygen species in the killing of the bacterial fish pathogen *Aeromonas salmonicida* by rainbow trout macrophages. *Fish & Shellfish Immunology*, 3: 119-129.

Takahara, S., Hamilton, H.B., Neel, J.V., Kobara, T.Y., Ogura, Y. and Nishimura, E.T. 1960. Hypocatalasemia: a new genetic carrier state. *The Journal of Clinical Investigation*, 39: 610-619.

Taw, N. 2006. Shrimp production in ASP system, CP Indonesia: Development of the technology from R&D to commercial production. *Aquaculture America.*

Wang, Y.B., Tian, Z.Q., Yao, J.T. and Li, W.F. 2008. Effect of probiotics, Enteroccus faecium, on tilapia (Oreochromis niloticus) growth performance and immune response. *Aquaculture*, 277: 203-207.

Zokaeifar, H., Balcázar, J.L., Saad, C.R., Kamarudin, M.S., Sijam, K., Arshad, A. and Nejat, N. 2012. Effects of Bacillus subtilis on the growth performance, digestive enzymes, immune gene expression and disease resistance of white shrimp, Litopenaeus vannamei. *Fish & shellfish immunology*, 33: 683-689.

CHAPTER 6

PROBIOTIC POTENTIAL OF *BACILLUS* STRAINS OF BIOFLOC SYSTEM ON GROWTH PERFORMANCE IN *PENAEUS VANNAMEI*

This study established the effectiveness of four different *Bacillus* sp., i.e., *Bacillus subtilis, B. megaterium, B. cereus* and *B. infantis* on growth performance, digestive enzyme activity and immune response in *Penaeus vannamei.* Experimental diets were identical in all the aspects except *Bacillus* incorporation: control (basal diet without probiotic) and four treatment diets T 1 (basal diet + *B. subtilis*), T 2 (basal diet + *B. megaterium*), T 3 (basal diet + *B. cereus*), T 4 (basal diet + *B. infantis*). Each diet was fed at 10% of animal body weight, four times in a day and reared for 6 weeks in triplicate tanks. Among the treatments, final weight (0.97 ± 0.06g), weight gain, specific growth rate and survival rate (64.66 ± 4.16%) were significantly greater in *B. subtilis* treated group. Interestingly, in digestive enzyme activities, protease (3.57 ± 0.42U/mg protein), lipase (747.83 ± 139.03 U/

mg protein) and amylase (7.53 ± 1.27U/mg protein) were considerably higher in *B. subtilis* supplemented diet fed group compared to other treatments. However, cellulase (246.83 ± 29.77U/mg protein) activity was much greater in the group that received *B. megaterium*. Activities of superoxide dismutase, phenol oxidase and catalase were analysed and the *B. subtilis* incorporated diet fed group exhibited better performance. Thus its concluded that *B.subtilis* improves digestive enzyme activities and growth performance and down-regulates the reactive oxygen species (ROS). This proved the versatility of *B. subtilis* as a probiotic in shrimp farming practices.

INTRODUCTION

Aquatic probiotics are considered as a safe additive, to provide health benefits to the cultured fish by enhancing growth and immunity, improving feed utilisation rate, increasing digestive enzyme activity, maintaining water quality, controlling diseases, modulating microbial colonisation in the intestine and pre-digestion of anti-nutritional factors present in the feed (Shen, Fu, Li, & Zhu, 2010; Selim & Reda, 2015; Amoah *et al.*, 2019). Therefore, probiotics have been considered as an effective alternative in place of the much criticised and banned antibiotics (Dawood, Koshio, & Esteban, 2018).

Among the probiotics, lactic acid bacteria (LAB) and *Bacillus* sp., have been proven to possess the capabilities of a long-lasting shelf life, production of antimicrobial substances as a secondary metabolite (non-pathogenic and non-toxic) and resistance to extreme pH and temperature. In addition to that, vast documentation of their beneficial effects in the aquaculture industry, encouraged their usage in several parts of the world (Zokaeifar *et al.*, 2014; Nimrat, Khaopong, Sangsong, Boonthai, & Vuthiphandchai, 2019). A growing number of studies in relation to probiotic administration revealed that, among the probiotics, *Bacillus* sp. has become more popular and widely used in shrimp aquaculture (Liu, Chiu, Ho, & Wang, 2009). *Bacillus* sp. a gram positive, rod shaped, endospore-forming bacteria enhances the antioxidant enzyme activity, involving in the expressions of immune and stress related genes (Liu *et al.*, 2009; Nayak, 2010; Buruiana, Profir & Vizireanu, 2014) and inhibits the virulent gene expression (quorum-sensing inhibition) (Musthafa,Saroja, Pandian, & Ravi, 2011; Subramenium, Swetha, Iyer, Balamurugan & Pandian, 2018).

B. subtilis, a probiotic spore forming bacterium, has been used in shrimp aquaculture as a growth and immune enhancer (Shen *et al.*, 2010). In the experimental diets, *B. subtilis* survived in greater numbers than *Lactobacillus acidophilus* at 4°C and 25°C (Aly, Ahmed, Ghareeb, & Mohamed, 2008). Similarly, a comparative study revealed that *B. subtilis* could provide better protection against vibriosis than *Lactobacillus plantarum* in larvae of sea bass (Touraki, Karamanlidou, Karavida & Chrysi, 2012). Likewise, inclusion of *Bacillus megaterium* in feed, as a probiotic, maintained the intestinal micro-flora balance which enhanced the growth and digestive enzyme activities in *Clarias* sp. (Afrilasari & Meryandini, 2016). It has been reported that, comparably, *B. infantis* has potential probiotic properties including, resistance to acid and bile salt, tendency to adhere to the intestinal wall and possess antagonistic activity against fish pathogens (Dharmaraj & Rajendren, 2014). Experimental inclusion of *Bacillus cereus* (10^8CFUg^{-1}), in the piglet rearing feed, enhanced the overall weight gain (Kirchgessner, Roth, Eidelsburger & Gedek, 1993). However, closely related bacterial strains might have different clinical effects and it is also emphasised that beneficial effects yielded by one bacterial strain cannot be assumed to occur with another bacterial strain (FAO, 2001). Similarly, antagonist activity shown by the probiotic bacteria vary according to microbial strains (Veron, Di Risio, Isla, & Torres, 2017). Therefore, it is necessary to identify the organisms at strain level for clinical studies.

Pacific white shrimp (*Penaeus vannamei*) serves as a widely cultured and economically important species in India as well as in the world, at a commercial level. It has the ability to grow in a wide range of salinities and temperatures and can be bred easily. It exhibits faster growth and higher survival rate even at a high stocking density, attains an attractive size in a short crop period and has high market demand. Another positive aspect is the availability of disease-free and disease resistant juveniles for culture (Chiu, Guu, Liu, Pan, & Cheng, 2007). However, in the recent times, a series of disease outbreaks have been encountered. Unlike, other aquatic animals, Pacific white shrimp lack in adaptive immune functions; instead they possess innate immune features to tackle such threats (Bachère *et al.,* 2004; Runsaeng, Kwankaew & Utarabhand, 2018).

Available evidence of close association and positive results between intestinal bacterial communities and host health in human (Nyangale, Farmer, Keller, Chernoff, & Gibson, 2014;

Majeed *et al.*, 2015), have given us much knowledge of probiotics. However, the information on the effects of probiotics in aquatic animals is yet to ascertain whether the supplementation of probiotics will positively impact the shrimps' intestinal microbial community and intestinal health as was reported in humans and other animals (Clemente, Ursell, Parfrey & Knight, 2012). The effect of probiotics on fish and shrimp has been reported (Wang, 2007; Krummenauer *et al.*, 2014; & García Bernal *et al.*, 2018). The present study was designed to find out the potential probiotic strain, among the *Bacillus* strains, which could be suggested as a future probiotic to enhance shrimp production.

METHODOLOGY

The study was conducted at the Advanced Research Farm Facility (ARFF), Tamil Nadu Dr. J. Jayalalithaa Fisheries University (TNJFU), Chennai, Tamil Nadu, India.

Bacterial spp Isolated from Bio Floc System

B. subtilis, *B. megaterium*, *B. cereus* and *B. infantis* were used as potential probiotics in this experiment. These bacterial species were isolated and identified from the biofloc system of GIFT tilapia culture at ARFF,Chennai. The four *Bacillus* sp. were received from Advanced Research Farm Facility (ARFF),Chennai. Under aseptic laboratory conditions, all the four *Bacillus* sp. were revived on MRS (de Man, Rogosa, and Sharpe) agar.The pure colonies were used for feed preparation. The *Bacillus* sp. were preserved at -20°C in 25% sterile glycerol stock solution for maintaining the pure stocks.

Probiotic Characterisation

Morphological and biochemical characterisation was performed in order to confirm the probiotic characteristics of those *Bacillus* sp. according to Bergey's Manual of Determinative Bacteriology (Schubert, Buchanan, & Gibbons, 1974). DNase activity was checked by using toluidine blue DNA agar base (Hi-Media, M6131-100G).

Determination of Antibacterial Activity Against Shrimp Pathogens

The antibacterial activity for all the four *Bacillus* sp. were performed, using agar well diffusion method, against the *Vibrio* pathogens (*Vibrio parahaemolyticus* and *V. alginolyticus*). Briefly, pathogenic bacteria, *V. parahaemolyticus* and *V. alginolyticus* were cultured in tryptone soy broth (TSB) supplemented with 1.5% of NaCl. After incubation at 30°C for 24 h, the bacterial density was approximately adjusted to 10^8 cfu ml^{-1}. Subsequently, pour plates were prepared using 20 ml of melted tryptone soy agar (TSA) with 10 µl of each pathogenic strain, in separate sterile petri-dishes. Then, the solidified medium was punched with a sterile steel borer of diameter 6.0 mm. Four *Bacillus* sp. were individually cultured in TSB and incubated at 30°C for 24 h using orbital shaker at 120 rpm. Then culture broth was centrifuged at 7000 rpm for 10 min at 4°C. 50 µl of supernatant was taken from the tube and slowly added into the previously prepared agar plates with corresponding pathogens. All the plates were incubated at 30°C for 24 h. After incubation, plates were observed and zone of inhibition was measured using a vernier caliper.

Experimental Diet Preparation

Four experimental diets were prepared, namely; Treatment-1 (basal diet + *B. subtilis* -T1), Treatment-2 (basal diet + *B. megaterium* -T2), Treatment-3 (basal diet + *B. cereus* -T3), Treatment-4 (basal diet + *B. infantis* -T4) and control (basal diet without probiotic).

Commercial shrimp feed (crude 35% protein) was used as a basal diet. While preparing the experimental diets, pure colonies of four *Bacillus* sp. were individually cultured in 250 ml flask with TSB broth in an orbital shaker (150 rpm) at 30°C for 24 h. The bacterial cells were harvested by centrifugation (7000 rpm for 5 minutes at 4°C). The bacterial pellets were then washed and re-suspended in sterile normal saline solution (NSS, 0.9% NaCl), and the optical density (OD) was adjusted to 1.000 at 540 nm. To the OD value adjusted culture, 0.5% of sodium alginate was added and 600 ml of bacterial inoculum was sprayed over 1kg of commercial feed. Then, the feed was transferred to an incubator, for drying at 37°C for 5 h, to obtain a required dosage of bacterial concentration in feed (10^8cfu g^{-1}). After incubation, feed was kept at 4°C and the cell viability was monitored. Feed was prepared, once in five days, to ensure the maximum survival of probiotic in the experimental diets.

Experimental Design and Animal Rearing

The study followed a completely randomised design, with one control and four treatments, receiving four different *Bacillus* sp.

at same concentration (10^8 cfu g^{-1}) each of which were triplicated. The trial was conducted for 6 weeks using low saline borewell water (2 ppt). Specific pathogen free healthy juvenile shrimp were procured from Aqua Nova Hatcheries (P) Ltd., Kancheepuram, Tamil Nadu. The procured seed were maintained in a Fiber Reinforced Plastic (FRP) tank of 1.5 tonne capacity, provided with an aeration system. The shrimp were acclimatised in the FRP tanks for a period of 10 days and fed with a commercial diet (crude protein – 35%), four times in a day (07:00, 11:00, 17:00 and 21:00) at 10% of its body weight.

Prior to stocking, (750 numbers; average weight- 0.02 ± 0.00 g) shrimps were starved for 24 h, and randomly weighed and stocked in 15 glass tanks (100 l capacity) at 50 shrimp per tank. The animals were fed with their respective treatment diets, four times in a day, at 10% of its body weight. Physico-chemical parameters of water such as dissolved oxygen and temperature were measured on daily basis and ammonia (NH_4^+/NH_3) – N, nitrite (NO_2) – N, nitrate (NO_3) – N, and pH once in a week, following the standard methods of APHA, (1995).

Bacteriological Analysis

The changes in total plate count (TPC), total *Vibrio* count (TVC) and total *Lactobacillus* count (TLC) of culture water and shrimp gastrointestinal tract (GIT) were examined on weekly basis. During each week, two shrimps from each tank (n=6/treatment) were randomly collected and their gastro-intestinal tract (GIT) was removed aseptically and macerated with saline for bacteriological plating. TPC, TVC, and TLC counts were estimated using tryptone soy agar (TSA, Hi-Media), thiosulphate-citrate-bile salt agar (TCBS, Hi-Media) and de Man, Rogosa, and Sharpe agar (MRS, Hi-Media), respectively (Vieira *et al.*, 2016).

Growth Performance

Weight gain of shrimp was monitored on weekly basis, using 10 shrimps from each tank (n=30/treatment). The growth parameters such as final weight, weight gain, specific growth rate (SGR), food conversion ratio (FCR) and survival rate were determined according to Zokaeifar *et al.* (2012): weight gain (WG; g) = Final weight (g) – Initial weight (g), Specific growth rate (SGR; %/ day) = (In [Final body weight (g)] - [Initial body weight (g)]/ Days of experiment) × 100, Food conversion ratio (FCR) = Total feed intake (g)/weight gain (g), Survival rate (%) = (Number of shrimp harvested / Number of shrimp stocked) × 100

Digestive Enzyme Analysis

At the end of feeding trial, three shrimps from each glass tank (n=9/treatment) were taken out and their GIT was collected for digestive enzyme analysis. Following the removal of ingested food items, the dissected GIT was mixed with chilled sucrose solution (0.25 M) and homogenised (5% homogenate) using poly pestles. The homogenate was then centrifuged at 5000 rpm for 10 min at 4°C. After centrifugation, supernatant was collected and used for further analysis. All the enzymatic activities were expressed in specific activity (U/ mg protein). Intestinal protein was estimated by following Lowry *et al.* (1951).

The protease activity of GIT was determined by following Drapeau (1976). A reaction mixture consisting of 2.5 ml of 1% casein (prepared in 0.01 N NaOH), 0.05 M tris-phosphate buffer (pH 7.8) and 0.1 ml tissue homogenate was incubated at 37°C for 15 min. Then 2.5 ml of 10% trichloroacetic acid (TCA) was added to start the reaction. Later, the whole content was filtered and the final OD was measured at 280 nm in UV spectrophotometer (UV-1800, Sl. No. 11635203671 CD, Shimadzu Corporation, Japan). In this assay, 1% casein was used as a substrate and a calibration curve was prepared using tyrosine as a standard solution. One unit of protease activity was defined as the number of micromoles of tyrosine released/min/mg of protein at 37°C.

The amylase activity of GIT samples were analysed according to Bernfeld (1955) using 1% of soluble starch as a substrate. Reaction was initiated by adding 1 ml of tissue homogenate into 1 ml of starch (prepared in 0.1 M phosphate buffer, pH 7.0) solution, in a test tube, and incubated for 15 min at 37°C. The reaction was stopped by the addition of 2 ml of 3, 5 dinitrosalicylic acid (DNSA). Later, the tubes were placed in a boiling water bath for 5 min. Then tubes were cooled and the volume was made up to 10 ml using distilled water. The intensity of the colour developed was recorded in a UV spectrophotometer at 560 nm. A calibration curve was prepared using maltose as a standard solution. One unit of amylase activity was defined as the number of micromoles of maltose released/min/mg of protein at 37°C.

The lipase activity of GIT samples were determined according to Cherry and Crandell (1932) using a stabilised emulsion of olive oil. A reaction mixture consisted of 3 ml of distilled water, 1 ml of tissue homogenate, 0.5 ml of phosphate buffer (0.1 M, pH 7.0), and 2 ml of olive oil emulsion, which was incubated for 24 h at 27°C. Subsequently, 3 ml of alcohol (95%) and two drops of phenolphthalein indicator were added and it was titrated against alkali (0.05 N NaOH), until the appearance of a permanent pink colour. At the same time, a control was prepared using an enzyme source and the enzyme activity was inactivated by keeping it in a boiling water bath for 15 min, prior to the addition of buffer and olive oil emulsion. A calibration curve was prepared using porcine type pancreatic lipase. One unit of lipase activity was expressed as the numbers of fatty acid released/min/mg of protein at 27°C.

Cellulase activity was determined by following Gonzalez-Peña, Anderson, Smith & Moreira (2002). The reaction mixture consisted of 1 ml of carboxymethyl cellulose (CMC) as a substrate, 1 ml of phosphate buffer (0.1 M, pH 6.8) and 1 ml of tissue homogenate in a test tube and it was incubated at 37°C for 1 h. After incubation, 0.5 ml of DNSA reagent was added to stop the reaction. The final absorbance was recorded in a spectrophotometer at 540 nm. A calibration curve was prepared using glucose. One unit of cellulase activity is defined as the amount of glucose released/min/mg protein.

Immunological Parameters

The haemolymph was drawn from the ventral sinus cavity of shrimp (two shrimps from each tank; n=6/treatment) using 1 ml syringe and 22 gauge needle. The collected haemolymph was allowed to clot for 2 h at room temperature (37°C). After centrifugation at 6000 rpm for 10 min, the serum was carefully transferred to 1 ml vials which was used for the analysis of serum protein and phenoloxidase (PO). The total serum protein was estimated by following the Lowry, Rosebrough, Farr & Randall (1951) method.

The phenoloxidase activity was measured by following the method of Gollas-Galván, Hernández-López & Vargas-Albores (1997) with slight modifications. PO activity was measured by recording the formation of dopachrome from L-diydroxyphenylalanine (L-DOPA, Sigma) spectrophotometrically at 490 nm for 60 min at 2 min intervals. The activity was determined by the increase of O.D/min under the assay conditions.

Antioxidant Enzymes

The catalase activity of hepato-pancreas samples was performed as described by Takahara *et al.* (1960). The supernatant (0.2 ml) was mixed with 1.2 ml of 0.05 M phosphate buffer (pH 7.0) in a reaction tube. Then, 1 ml of 0.03 M H_2O_2 (prepared in phosphate buffer) was added to the reaction mixture. The OD value was recorded at 240 nm for 2 min at 30 sec interval. Simultaneously, the blank was run with 1 ml of distilled water instead of H_2O_2. Catalase activity was expressed as μ moles of H_2O_2 decomposed/min/mg protein.

Superoxide dismutase (SOD) activity of hepato-pancreas samples was performed by following the method of Mishra and Fridovich (1972) with slight modifications. The reaction mixture was prepared using 50 µl of tissue homogenate, 0.5 ml of epinephrine (0.03 M) and 1.5 ml of phosphate buffer (pH 7.4). The mixture was well mixed and the changes in OD was recorded at 480 nm for 2 min at 30 sec interval in UV spectrophotometer. One unit of SOD activity was determined by the amount of protein required to give 50% inhibition of epinephrine auto-oxidation.

The collected data on growth parameter, bacteriological analysis of culture water and digestive tract of cultured shrimp, enzymatic activities and immunological parameters were analysed in one-way analysis of variance (ANOVA) using SPSS version 16.0. Duncan's multiple range test was used for post hoc comparison of mean values among the treatments and the statistical significance of the test was set at $P<0.05$.

RESULTS AND CONCLUSIONS

Probiotic Characterisation

Probiotic properties of the four *Bacillus* sp. were characterised and presented in Table 1. All the four *Bacillus* sp. were positive in catalase activity, casein hydrolysis and starch hydrolysis and were negative in DNase activity. In the amino acid (AA) utilisation test, serine was the only amino acid utilised by all the *Bacillus* sp. and histidine was used only by *B. cereus.*

Table 1: Morphological and biochemical characteristics of four different strains of *Bacillus* sp. isolated from the biofloc system of GIFT tilapia culture

Test items	*B. subtilis*	*B. megaterium*	*B. cereus*	*B. infantis*
Gram staining	Positive	Positive	Positive	Positive
Cell morphology	Rod	Rod	Rod	Rod
Spores-forming	+	+	+	+
Motility	+	+	+	+
Catalase	+	+	+	+
DNase activity	–	–	–	–
Proline	–	–	–	–
Lysine	–	–	–	–
Ornithine	–	–	–	–
Serine	+	+	+	+
Histidine	–	–	+	–
Arginine	–	–	–	–
Casein hydrolysis	+	+	+	+
Starch hydrolysis	+	+	+	+
Gelatin hydrolysis	+	+	+	
Citrate utilization	–	–	–	–
Growth at 5% NaCl	+	+	+	+

Growth (+) or No growth (–) after incubation for 24 hours at 30°C.

Antibacterial Activity of *Bacillus* sp. Against Shrimp Pathogens

In the present study, *B. subtilis* and *B. cereus* exhibited maximum inhibition zone of (14, 15 and 13, 16 mm) against *V. parahaemolyticus* and *V. alginolyticus*, respectively, followed by *B. infantis* (Table 2). No inhibition zone was observed in *B. megaterium* against shrimp pathogenic species.

Table 2: Determination of antagonism of selected strains of *Bacillus* sp. against shrimp pathogens (*Vibrio parahaemolyticus* and *Vibrio alginolyticus*)

Strains	VP (mm)	VA (mm)
B. subtilis	14	15
B. megaterium	-	-
B. cereus	13	16
B. infantis	12	12

VP, *V. parahaemolyticus*; VA, *V. alginolyticus*

Bacteriological Analysis

A substantial difference (*P*< 0.05) in total plate count was recorded in culture water, except the first and third week (Table 3). The total plate count was observed to be much lower in all the treatments, except in control ($3.30 \pm 0.23 \times 10^5$ cfu/ml) during the last week. The difference (*P*< 0.05) in total *Vibrio* count was recorded among the treatments, except second and third week. The total *Vibrio* count in control was observed to be much higher ($1.76 \pm 0.29 \times 10^5$ cfu/ml) in the last week. A large difference in the total *Lactobacillus* count (*P*< 0.05) was observed in third, fifth and last week. Shrimp group fed with diet incorporated with *B. subtilis* displayed significantly higher total *Lactobacillus* count in culture water ($6.66 \pm 1.33 \times 10^3$ cfu/ml). In contrast *Lactobacillus* count was not observed in control group in the last week.

In intestinal tract, major difference (*P*< 0.05) in total plate count was recorded in fifth and last week. Significantly higher total plate count was observed in control group ($2.70 \pm 0.64 \times 10^7$ cfu/ g) in second week. The study did not find any

Table 3: Bacterial microbiota in the culture water and intestinal tract of *P. vannamei* fed with control diet and diet supplemented with different *Bacillus* strains

Time	Group	Microbial evaluation of the culture water (cfu/ml)			Microbial evaluation of the intestinal tract (cfu/g)		
(Week)		TPC	TVC	TLC	TPC	TVC	TLC
1	Control	$4.50 \pm 0.50 \times 10^{3}$	$1.96 \pm 0.06 \times 10^{4(a)}$	$5.00 \pm 0.25 \times 10^{1}$	$3.83 \pm 0.20 \times 10^{5}$	$4.23 \pm 0.22 \times 10^{4}$	$1.30 \pm 0.86 \times 10^{2}$
	T1	$3.30 \pm 1.20 \times 10^{4}$	$8.33 \pm 1.45 \times 10^{3(b)}$	$9.00 \pm 0.41 \times 10^{2}$	$1.30 \pm 0.25 \times 10^{5}$	$1.90 \pm 0.12 \times 10^{4}$	$5.63 \pm 0.91 \times 10^{2}$
	T2	$2.00 \pm 0.57 \times 10^{4}$	$8.66 \pm 0.37 \times 10^{3(b)}$	$2.70 \pm 1.19 \times 10^{3}$	$7.66 \pm 0.33 \times 10^{4}$	$4.03 \pm 0.70 \times 10^{4}$	$4.06 \pm 0.20 \times 10^{2}$
	T3	$4.30 \pm 0.20 \times 10^{4}$	$7.33 \pm 1.45 \times 10^{3(b)}$	$3.70 \pm 1.30 \times 10^{3}$	$1.30 \pm 0.26 \times 10^{5}$	$4.33 \pm 0.16 \times 10^{4}$	$8.96 \pm 0.65 \times 10^{2}$
	T4	$1.60 \pm 0.30 \times 10^{4}$	$7.83 \pm 1.74 \times 10^{3(b)}$	$1.60 \pm 0.60 \times 10^{3}$	$4.16 \pm 0.48 \times 10^{5}$	$2.83 \pm 0.60 \times 10^{4}$	$1.47 \pm 0.11 \times 10^{3}$
	P value	0.382	0.008	0.391	0.394	0.535	0.515
2	Control	$1.30 \pm 0.05 \times 10^{5(a)}$	$1.96 \pm 0.06 \times 10^{4}$	$1.00 \pm 0.01 \times 10^{2(b)}$	$2.70 \pm 0.64 \times 10^{7(a)}$	$3.56 \pm 1.58 \times 10^{6}$	$1.70 \pm 0.11 \times 10^{2(b)}$
	T1	$4.00 \pm 0.57 \times 10^{4(b)}$	$1.20 \pm 0.56 \times 10^{4}$	$1.40 \pm 0.90 \times 10^{4(a)}$	$1.20 \pm 0.58 \times 10^{7(ab)}$	$1.40 \pm 0.65 \times 10^{6}$	$2.90 \pm 0.15 \times 10^{2(a)}$
	T2	$6.66 \pm 1.66 \times 10^{4(b)}$	$8.00 \pm 0.36 \times 10^{4}$	$5.00 \pm 0.20 \times 10^{3(ab)}$	$2.03 \pm 0.89 \times 10^{7(ab)}$	$4.20 \pm 0.23 \times 10^{6}$	$1.66 \pm 0.33 \times 10^{2(a)}$
	T3	$1.10 \pm 0.20 \times 10^{5(a)}$	$1.30 \pm 0.32 \times 10^{4}$	$3.00 \pm 1.15 \times 10^{3(ab)}$	$5.46 \pm 0.47 \times 10^{6(b)}$	$3.93 \pm 1.14 \times 10^{6}$	$2.33 \pm 0.33 \times 10^{2(a)}$
	T4	$1.26 \pm 0.12 \times 10^{5(a)}$	$2.66 \pm 0.21 \times 10^{4}$	$2.33 \pm 0.88 \times 10^{3(ab)}$	$1.36 \pm 0.60 \times 10^{7(ab)}$	$4.53 \pm 0.58 \times 10^{6}$	$5.00 \pm 0.20 \times 10^{2(a)}$
	P value	0.000	0.781	0.203	0.236	0.138	0.389
3	Control	$9.00 \pm 1.15 \times 10^{4}$	$3.65 \pm 0.27 \times 10^{4(b)}$	$1.33 \pm 0.33 \times 10^{2(c)}$	$1.10 \pm 0.65 \times 10^{7}$	$1.36 \pm 0.96 \times 10^{6(a)}$	$5.00 \pm 0.23 \times 10^{2(c)}$
	T1	$5.33 \pm 1.45 \times 10^{4}$	$3.33 \pm 0.23 \times 10^{3(b)}$	$1.10 \pm 0.49 \times 10^{3(bc)}$	$1.36 \pm 0.82 \times 10^{6}$	$1.20 \pm 0.41 \times 10^{5(c)}$	$1.76 \pm 0.78 \times 10^{4(a)}$
	T2	$6.33 \pm 0.33 \times 10^{4}$	$5.00 \pm 0.30 \times 10^{3(b)}$	$1.13 \pm 0.46 \times 10^{3(bc)}$	$1.60 \pm 0.80 \times 10^{6}$	$5.70 \pm 0.96 \times 10^{5(b)}$	$4.66 \pm 0.27 \times 10^{3(b)}$
	T3	$5.66 \pm 1.76 \times 10^{4}$	$1.00 \pm 0.04 \times 10^{5(a)}$	$3.33 \pm 0.88 \times 10^{3(b)}$	$5.33 \pm 0.33 \times 10^{6}$	$3.96 \pm 0.70 \times 10^{5(b)}$	$1.46 \pm 0.12 \times 10^{4(a)}$
	T4	$5.13 \pm 1.16 \times 10^{4}$	$1.03 \pm 0.54 \times 10^{3(b)}$	$5.33 \pm 1.76 \times 10^{3(a)}$	$3.73 \pm 0.46 \times 10^{6}$	$3.96 \pm 0.89 \times 10^{5(b)}$	$1.56 \pm 0.12 \times 10^{4(a)}$
	P value	0.304	0.300	0.017	0.275	0.448	0.707

[Table Contd.

Contd. Table]

Time (Week)	Group	Microbial evaluation of the culture water (cfu/ml)			Microbial evaluation of the intestinal tract (cfu/g)		
		TPC	TVC	TLC	TPC	TVC	TLC
4	Control	$9.00 \pm 2.30 \times 10^{4(a)}$	$1.20 \pm 0.20 \times 10^{5(a)}$	$3.00 \pm 0.57 \times 10^{2(b)}$	$2.13 \pm 0.44 \times 10^{6(ab)}$	$5.50 \pm 0.32 \times 10^{5(a)}$	$3.10 \pm 0.19 \times 10^{2(b)}$
	T1	$3.33 \pm 0.88 \times 10^{4(b)}$	$4.36 \pm 1.73 \times 10^{3(b)}$	$1.66 \pm 0.66 \times 10^{3(ab)}$	$7.66 \pm 0.48 \times 10^{5(b)}$	$1.10 \pm 0.10 \times 10^{5(b)}$	$2.73 \pm 0.81 \times 10^{4(ab)}$
	T2	$2.00 \pm 0.57 \times 10^{4(b)}$	$3.66 \pm 0.26 \times 10^{3(b)}$	$1.33 \pm 0.33 \times 10^{3(ab)}$	$5.66 \pm 0.29 \times 10^{5(b)}$	$1.10 \pm 0.23 \times 10^{5(b)}$	$4.36 \pm 1.73 \times 10^{4(a)}$
	T3	$3.00 \pm 0.00 \times 10^{4(b)}$	$7.00 \pm 1.52 \times 10^{3(b)}$	$2.00 \pm 0.57 \times 10^{3(a)}$	$1.20 \pm 0.11 \times 10^{6(b)}$	$2.80 \pm 0.16 \times 10^{5(b)}$	$2.73 \pm 1.39 \times 10^{4(ab)}$
	T4	$5.33 \pm 0.88 \times 10^{4(b)}$	$3.00 \pm 1.52 \times 10^{3(b)}$	$1.33 \pm 0.33 \times 10^{3(ab)}$	$3.70 \pm 0.14 \times 10^{6(a)}$	$2.30 \pm 0.13 \times 10^{5(b)}$	$1.43 \pm 0.80 \times 10^{4(ab)}$
	P value	0.006	0.000	0.177	0.055	0.380	0.162
5	Control	$2.10 \pm 0.30 \times 10^{5(a)}$	$1.20 \pm 0.20 \times 10^{5(a)}$	$3.33 \pm 0.33 \times 10^{2(b)}$	$7.33 \pm 2.33 \times 10^{6(a)}$	$5.73 \pm 0.28 \times 10^{5(a)}$	$6.66 \pm 0.88 \times 10^{2(b)}$
	T1	$6.00 \pm 0.26 \times 10^{4(bc)}$	$7.66 \pm 1.76 \times 10^{3(b)}$	$2.33 \pm 0.33 \times 10^{3(a)}$	$1.00 \pm 0.15 \times 10^{6(b)}$	$5.63 \pm 0.23 \times 10^{4(ab)}$	$4.66 \pm 0.36 \times 10^{4(a)}$
	T2	$4.20 \pm 1.94 \times 10^{4(c)}$	$5.00 \pm 0.23 \times 10^{3(b)}$	$2.00 \pm 0.02 \times 10^{3(a)}$	$1.33 \pm 0.33 \times 10^{6(b)}$	$2.53 \pm 0.22 \times 10^{5(ab)}$	$4.33 \pm 0.66 \times 10^{4(a)}$
	T3	$1.56 \pm 0.37 \times 10^{5(ab)}$	$7.00 \pm 1.52 \times 10^{3(b)}$	$2.33 \pm 0.33 \times 10^{3(a)}$	$1.90 \pm 0.58 \times 10^{6(b)}$	$4.66 \pm 1.45 \times 10^{4(b)}$	$2.50 \pm 0.50 \times 10^{4(a)}$
	T4	$1.43 \pm 0.49 \times 10^{4(c)}$	$5.33 \pm 0.23 \times 10^{3(b)}$	$2.00 \pm 0.57 \times 10^{3(a)}$	$3.16 \pm 1.92 \times 10^{5(b)}$	$1.12 \pm 0.94 \times 10^{5(ab)}$	$7.66 \pm 1.61 \times 10^{4(a)}$
	P value	0.015	0.000	0.005	0.005	0.196	0.621
Last week	Control	$3.30 \pm 0.23 \times 10^{5(a)}$	$1.76 \pm 0.29 \times 10^{5(a)}$	$0.00^{(c)}$	$1.90 \pm 0.60 \times 10^{7(a)}$	$7.00 \pm 0.23 \times 10^{5(a)}$	$8.33 \pm 2.40 \times 10^{2(c)}$
	T1	$3.33 \pm 0.88 \times 10^{3(b)}$	$4.40 \pm 0.28 \times 10^{3(b)}$	$6.66 \pm 1.33 \times 10^{3(a)}$	$1.66 \pm 0.20 \times 10^{5(b)}$	$1.93 \pm 0.10 \times 10^{4(b)}$	$5.33 \pm 0.88 \times 10^{4(b)}$
	T2	$2.00 \pm 0.57 \times 10^{3(b)}$	$1.20 \pm 0.43 \times 10^{4(b)}$	$2.33 \pm 0.88 \times 10^{3(b)}$	$9.33 \pm 0.88 \times 10^{5(b)}$	$2.56 \pm 0.17 \times 10^{4(b)}$	$4.33 \pm 1.45 \times 10^{4(ab)}$
	T3	$3.00 \pm 0.01 \times 10^{3(b)}$	$1.43 \pm 0.20 \times 10^{4(b)}$	$2.23 \pm 0.76 \times 10^{3(b)}$	$6.36 \pm 0.30 \times 10^{5(b)}$	$3.46 \pm 0.19 \times 10^{4(b)}$	$4.33 \pm 0.88 \times 10^{4(ab)}$
	T4	$3.23 \pm 1.29 \times 10^{4(b)}$	$1.00 \pm 0.37 \times 10^{4(b)}$	$1.66 \pm 0.66 \times 10^{3(b)}$	$3.76 \pm 0.21 \times 10^{5(b)}$	$1.06 \pm 0.54 \times 10^{4(b)}$	$8.66 \pm 1.24 \times 10^{4(a)}$
	P value	0.000	0.000	0.006	0.001	0.001	0.006

In each column values (mean ± SD, n=6) with different superscripts differ significantly ($P < 0.05$).

noteworthy difference ($P > 0.05$) in total *Vibrio* count in the shrimp's intestinal tract, except in last week. Interestingly, in last week, higher total *Vibrio* count was recorded in control ($7.00 \pm 0.23 \times 10^5$cfu/g). There was not much of a difference ($P > 0.05$) in total *Lactobacillus* count, except last week. In *Lactobacillus* count, among the different treatments, significantly higher total *Lactobacillus* count was recorded in T4 group ($8.66 \pm 1.24 \times 10^4$ cfu/g) in the last week.

Growth Performance and Water Quality Parameters

The present study found a considerable difference ($P < 0.05$) in the final weight, weight gain and specific growth rate (Table 4). Significantly higher and lower growth performance of final weight (0.97 ± 0.06 & 0.64 ± 0.01 g), weight gain (0.94 ± 0.06 & 0.61 ± 0.01 g) and specific growth rate (7.88 ± 0.18 & 6.96 ± 0.04 %/ day) were recorded in T1 and control, respectively. Similarly, a remarkably higher survival rate was noticed in T1 (64.66 ± 4.16%). Further, the shrimp fed with *B. subtilis* and *B. megaterium* incorporated diets exhibited better FCR (1.04 ± 0.05 & 1.13 ± 0.06). In contrast to this, the control group showed poor feed utilisation, in terms of FCR (1.31 ± 0.03).

All the water quality parameters within the acceptable ranges; DO (5.0 ± 0.5 mg/l), temperature (30 ± 1.0°C), pH (7.8 – 8.3), ammonia-N (< 0.1 ± 0.01 mg/ l), nitrite-N (< 0.05 ± 0.001 mg/ l) and nitrate-N (< 3.0 ± 1.0 mg/l) were recorded during the experimental period.

Table 4: Effect of supplementation of different *Bacillus* strains on growth performance and survival of *P. vannamei*

Treatments	Initial weight (g)	Final weight (g)	Weight gain (g)	SGR (%/day)	FCR	Survival (%)
Control	0.02 ± 0.00	0.64 ± 0.01^{c}	0.61 ± 0.01^{c}	6.96 ± 0.04^{c}	1.31 ± 0.03^{a}	51.33 ± 2.30^{c}
T1	0.02 ± 0.00	0.97 ± 0.06^{a}	0.94 ± 0.06^{a}	7.88 ± 0.18^{a}	1.04 ± 0.05^{b}	64.66 ± 4.16^{a}
T2	0.02 ± 0.00	0.72 ± 0.00^{bc}	0.69 ± 0.00^{bc}	7.25 ± 0.02^{b}	1.13 ± 0.06^{b}	56.00 ± 4.00^{bc}
T3	0.02 ± 0.00	0.81 ± 0.06^{b}	0.78 ± 0.06^{b}	7.46 ± 0.15^{b}	1.22 ± 0.08^{ab}	60.66 ± 3.05^{ab}
T4	0.02 ± 0.00	0.78 ± 0.03^{b}	0.75 ± 0.03^{b}	7.40 ± 0.20^{b}	1.16 ± 0.04^{ab}	60.66 ± 6.42^{ab}
P value	0.587	0.002	0.002	0.002	0.065	0.006

In each column, values (mean ± SD, n=30) with different superscripts differ significantly ($P < 0.05$)

Table 5: Effect of supplementation of different strains of *Bacillus* sp. dietary probiotic on immune performance of *P. vannamei*

Treatments	Serum protein (g/dl)	SOD (U/mg protein)	Catalase (U/mgprotein)	PO (U/mg protein)
Control	3.60 ± 0.26	287.44 ± 34.55^{a}	8.92 ± 0.59^{a}	20.24 ± 2.77^{a}
T1	3.83 ± 0.36	129.72 ± 28.04^{c}	3.67 ± 0.98^{b}	9.60 ± 0.61^{b}
T2	3.52 ± 0.57	232.51 ± 16.79^{ab}	5.17 ± 0.22^{b}	22.83 ± 4.26^{a}
T3	3.76 ± 0.39	217.41 ± 17.58^{ab}	6.34 ± 1.46^{b}	21.65 ± 2.46^{a}
T4	4.16 ± 0.55	144.29 ± 17.99^{bc}	3.84 ± 1.12^{b}	8.93 ± 0.46^{b}
P value	0.883	0.018	0.007	0.013

In each column, values (mean ± SD, n=6) with different superscripts differ significantly ($P < 0.05$)

Digestive Enzyme Activity

A significant difference ($P < 0.05$) in amylase activity was observed among the shrimp groups fed with different probiotic sp (Fig.1). Among these, a higher amylase activity was recorded in *B. subtilis* supplemented group (T1) (7.53 ± 1.27 U/mg protein) and lower amylase activity in the control group (1.59 ± 0.28 U/mg protein). There was no substantial difference in lipase activity reported in the *Bacillus* sp. fed groups (Fig.2). However, a lower lipase activity was recorded in control group (210.23 ± 26.33 U/mg protein). The protease activity differed widely ($P < 0.05$) among the different treatment groups (Fig. 3). A notably higher protease activity was noticed in *B. subtilis* (T1) fed group (1.89 ± 0.41U/mg protein) followed by *B. megaterium* (T2) fed group (2.23 ± 0.03U/mg protein). In cellulase activity, a considerable difference ($P < 0.05$) was observed among the treatments. A much higher cellulase activity was recorded in T2 group (246.83 ± 29.77 U/mg protein) (Fig. 4).

Immunological Parameters

There was no major difference in serum protein levels of different treatment groups. The phenol oxidase was significantly lower in T4 (8.93 ± 0.46U/mg protein) followed by T1 (9.60 ± 0.61 U/mg protein) compared to other treatment groups.

Antioxidant Enzymes

In SOD, a vital difference ($P < 0.05$) was observed among the different treatment groups. A higher SOD was recorded in

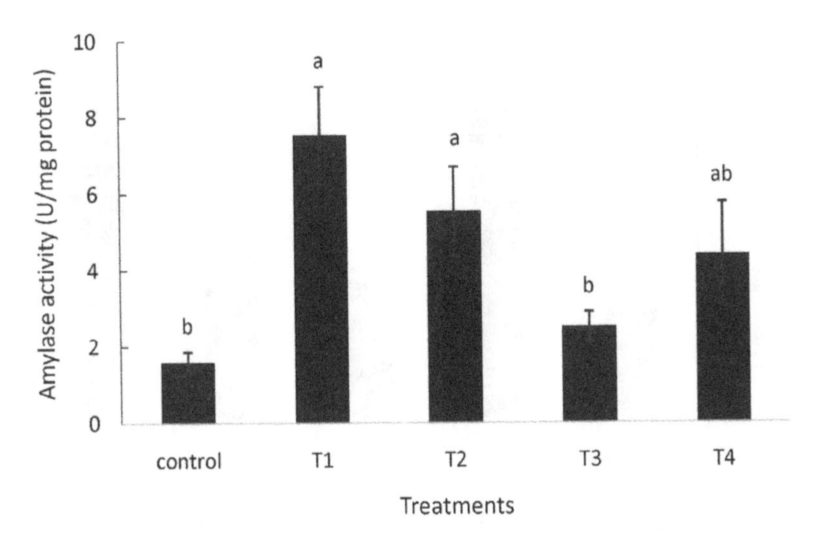

Fig. 1. Specific activity of intestinal amylase (U/mg protein) from shrimp fed with different strains of *Bacillus* incorporated diets. Bars with different superscripts differ significantly at $P < 0.05$.

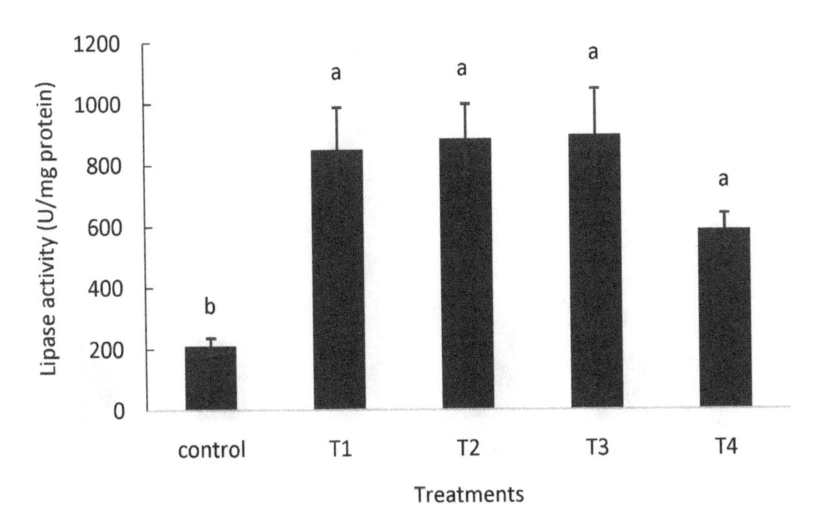

Fig. 2. Specific activity of intestinal lipase (U/mg protein) from shrimp fed with different strains of *Bacillus* incorporated diets. Bars with different superscripts differ significantly at $P < 0.05$.

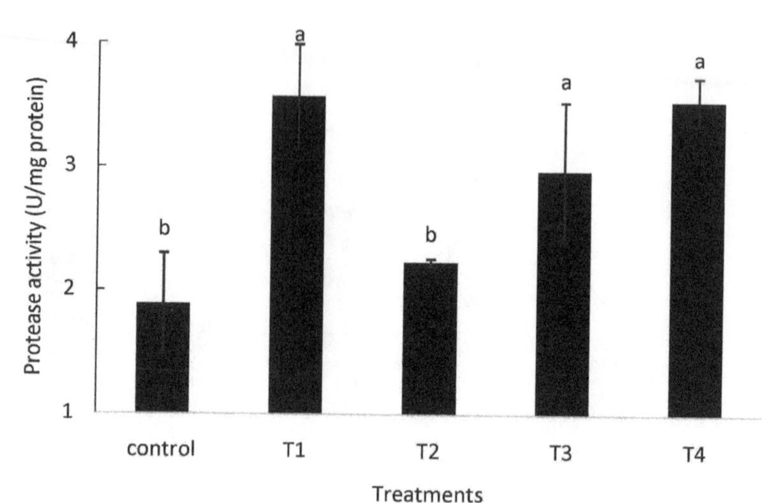

Fig. 3. Specific activity of intestinal protease (U/mg protein) from shrimp fed with different strains of *Bacillus* incorporated diets. Bars with different superscripts differ significantly at $P < 0.05$.

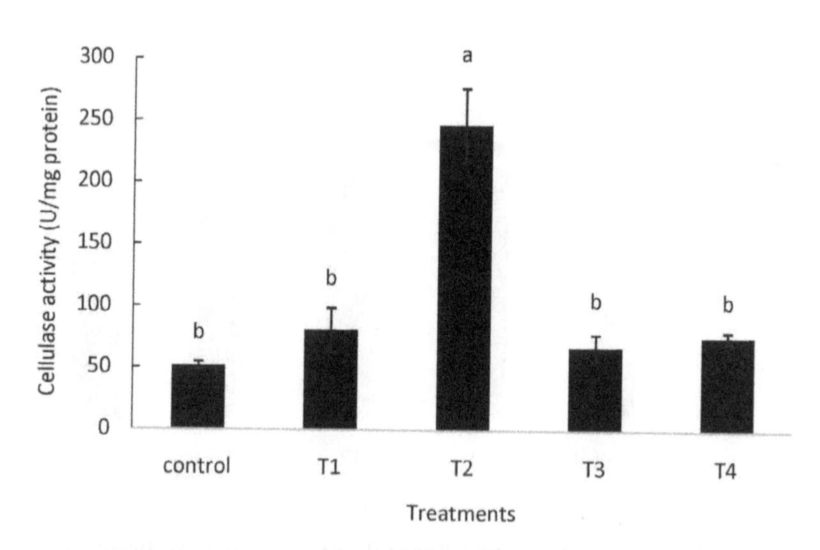

Fig. 4. Specific activity of intestinal cellulase (U/mg protein) from shrimp fed with different strains of *Bacillus* incorporated diets. Bars with different superscripts differ significantly at $P < 0.05$.

control (287.44 ± 34.55 U/mg protein) followed by T2 (232.51 ± 16.79 U/mg protein) and lower superoxide dismutase activity was observed in T1 (129.72 ± 28.04 U mg/ protein). The catalase activity, in the control group, was significantly higher than the other treatment groups.

Almost all the practices have been tried in aquaculture sector, to challenge pathogenic intruders, except the eco-friendly practice of probiotic supplementation. Many studies have suggested the usage of probiotics as a part of aquaculture practice (Liu *et al.*, 2009; Buruiana *et al.*, 2014 & Zokaeifar *et al.*, 2014). It has been well documented that *Bacillus* sp. are able to produce ranges of extracellular substances and antimicrobial peptides, which in turn improve growth, digestion and absorption of feed, health, immunity and survival against pathogenic microorganisms (Zokaeifar *et al.*, 2012; Wang, Hu, Chiu,& Liu, 2019).

The effective concentration of single probiotic strain in a diet, to improve the growth performance and health status of aquatic animal is 10^8 cfu g^{-1} (Zokaeifar *et al.*, 2012; Amoah *et al.*, 2019). Therefore, in the present study, concentration of 10^8 cfu g^{-1} was supplemented to the different treatment diets with their respective probiotic bacteria. In aquaculture, most of the studies on probiotics have been carried out using a single probiotic strain or compared two species. Notably, less is known about the potential effects of different species of *Bacillus* in a particular sphere of work.

DNase is an enzyme holding the potential virulence factors (bacterial innate immune avoidance mechanisms), such as subdual effect on bactericidal activity of macrophage and TLRs (toll like receptors) - mediated innate immune response. The findings of the present study illustrated that none of the *Bacillus* sp. were able to produce DNase which confirmed the probiotic potentiality of the selected species.

In the selected strain, *B. subtilis* and *B. cereus* showed inhibitory activity against *V. parahaemolyticus* and *V. alginolyticus* with an average inhibition zone of 14, 15 and 13, 16 mm, respectively. Gullian, Thompson & Rodriguez. (2004) reported that, species of *Bacillus* P64 had a maximum inhibition effect against *V. harveyi* S2 with 15.3 mm of inhibition zone. Similar results were reported by Luis-Villasenor, Macías-Rodríguez, Gómez-Gil, Ascencio-Valle & Campa-Córdova (2011) using *Bacillus* spp. against *V. parahaemolyticus* (CAIM 170) and *V. harveyi* (CAIM 1793) with an inhibition zone of 11 and 17.5 mm, respectively. Generally, *Bacillus* sp. are well known for its ability to produce proteinaceous substances, which encompass antibacterial compounds such as enzymes and bacteriocins or bacteriocin like inhibitory substances (BLIS) (Zokaeifar *et al.*, 2012). These may exert a bactericidal or bacteriostatic effect on ranges of Gram-positive and Gram-negative pathogens (Schulz *et al.*, 2005). This could be the reason for the inhibitory activity of *Bacillus* sp. against *Vibrio* pathogen.

Recently, the intestinal microflora gained more attention due to its crucial role on shaping structural integrity of intestinal features (Amoah *et al.*, 2019). It has also been confirmed that probiotic supplementation positively alters the gut microflora by producing antimicrobial peptides to override the growth of other organisms (Kar and Ghosh, 2008). The first line of action of probiotic on host is competitive exclusion- a mechanism by which the probiotic creates a hostile environment for pathogen colonisation in the GIT of the host animal. Interestingly, in the present study, successful colonisation of different *Bacillus* sp. and considerable reduction of total Vibrio count in the GIT of fed animal was noticed. This proved the successful competitive exclusion of pathogen by the supplemented Bacillus sp. from the shrimp GIT. Similar results were obtained by Zokaeifar *et al.*(2012) using *B. subtilis* species L10 and G1 as probiotics. It has been reported that probiotic, *Bacillus* can multiply in the digestive tract of host animal and perform, similar to oral vaccines (Interaminense *et al.*, 2018).

In shrimp culture, the emerging threats, namely new disease outbreaks and stressful conditions, could be reduced and higher production, with improved weight gain, SGR, FCR, digestion and nutrient absorption, can be achieved using probiotics (Chai, Song, Chen, Xu & Huang, 2016). In the present study, significantly improved growth performance, in terms of weight gain, SGR, FCR and better survival, was observed in *B. subtilis* incorporated diet than the other treatment groups. Similarly, *P. vannamei* fed with *B. subtilis* incorporated diet showed increased growth performance than control and *B. megaterium* supplemented groups (Olmos, Ochoa, Paniagua-Michel, & Contreras, 2011).

The present study noticed a significant increase in digestive enzyme activity of probiotic supplemented diet fed groups. In addition to this, good appetite behaviour was observed in all the treatment groups, except control group. Previous reports suggest that incorporation of *Bacillus* sp. or probiotics enhances the specific activity of lipase, amylase and protease in the digestive tract (Ziaei-Nejad, 2006; Amoah *et al.*, 2019). The reason for elevated digestive enzyme activities in the supplemented groups is due to the production of exogenous digestive enzymes by the probiotic bacterial species, after successful colonisation in GIT, which subsequently induce the endogenous production of digestive enzymes (Wang, 2007 & Liu *et al.*, 2009). Overall, *B. subtilis* fed group exhibited increased digestive enzyme activities, except cellulase, which can be further correlated with increased growth of the shrimp in that group. Similarly, *P. vannamei* juveniles fed with *B. subtilis* in the diet displayed increased digestive enzyme activities, protease and amylase and better growth performance (Zokaeifar *et al.*, 2014).

Higher vertebrates, while encountering a foreign substance, activate acquired immune responses, whereas invertebrates, especially shrimps, use innate immune responses (Tseng *et al.*, 2009). To further strengthen the simple immune system of shrimp, probiotics are intended to be used in aquaculture as a sustainable practice to encounter the bacterial or viral intruders. Serum protein concentration indicates the immune status of fish. Higher concentration directly correlates with stronger immune response. Serum proteins (albumin and globulin) are novel components necessary for the efficient functioning of the immune system (Yengkokpam *et al.*, 2016). The present study did not find any significant difference in total serum protein among the different treatment groups. In contrast, Ferreira *et al.* (2015) reported increased serum protein in shrimp fed with *Bacillus* sp. than compared to control. However, no major difference was observed in total serum protein of *P. vannamei* fed with *Bacillus* sp. (Gullian *et al.*, 2004; Liu *et al.*, 2009).

The present study revealed that PO activity was considerably higher in all the treated groups, except those fed with *B. infantis* and *B. subtilis* incorporated diet. Ferreira *et al.* (2015) noticed lower PO activity in *P. vannamei* treated with *Bacillus* spp. supplemented diet. In contrary to that, Gullian *et al.* (2004) noticed a much higher PO activity in shrimp fed with probiotics. A striking increase in PO activity of groups treated with probiotics has been attributed only after challenging the shrimp with *Vibrio* spp. or any other pathogens (Chiu *et al.*, 2007).

The superoxide dismutase activity was significantly higher in control and lower in *B. subtilis* supplemented diet fed groups. In the process of metabolism, aerobic organisms continuously produce endogenous reactive oxygen species (ROS), which in turn cause potential cytotoxic problems to the host (Munoz *et al.*, 2000). Therefore, in the present study, to avoid the cytotoxic effect by nullifying the excessive ROS (Shen *et al.*, 2010), SOD level might have increased in control, T2, T3 groups.

Catalase (CAT), another antioxidant enzyme, protects organisms against the oxidative stress by decomposing hydrogen peroxide (Tavares-Sánchez *et al.*, 2004). The present study found significantly lower catalase activity in different *Bacillus* species treated groups compared to control group. The result indicated lower generation of H_2O_2 in treatment groups, which received different *Bacillus* species as a feed supplement. Castex, Lemaire, Wabete & Chim (2010) reported decreased CAT and lower free radical generation when Pacific blue shrimp, (*Litopenaeus stylirostris*) was supplemented with effective microorganisms (EM) in diet.

The presence of too many toxic molecules, such as quinones, hemiquinones and free radicals (ROIs and NIRs), in the absence of infection, leads to unnecessary energy burden to the shrimp, which in turn negatively affects the tissues (Munoz *et al.*, 2000). While considering the above statement, findings of the present study revealed that *B. subtilis* could be a better probiotic strain, among the other *Bacillus* sp. by reducing the superoxide dismutase, and catalase activities.

Shrimp fed with *B. subtilis* incorporated diet showed potential competitive exclusion, better growth performance and improved digestive enzyme activities, with optimal anti-oxidant defence performance. This proved that *B. subtilis* would be a better single probiotic strain for ensuring complete biosecurity of shrimp farming practices in future.

ACKNOWLEDGEMENT

This research has been supported by grants from Department of Biotechnology (DBT), Govt of India, New Delhi (Project code: 507; 2016 -2019; PI: Dr.S.Felix).

REFERENCES

Afrilasari, W. and Meryandini, A. 2016. Effect of probiotic *Bacillus megaterium* PTB 1.4 on the population of intestinal microflora, digestive enzyme activity and the growth of Catfish (*Clarias* sp.). *Hayati Journal of Biosciences*, 23(4): 168-172. https://doi.org/10.1016/j.hjb.2016.12.005

Aly, S.M., Ahmed, Y.A.G., Ghareeb, A.A.A. and Mohamed, M.F. 2008. Studies on *Bacillus subtilis* and *Lactobacillus acidophilus*, as potential probiotics, on the immune response and resistance of Tilapia nilotica (*Oreochromis niloticus*) to challenge infections. *Fish & Shellfish Immunology*, 25(1-2): 128-136. https://doi.org/10.1016/j.fsi.2008.03.013

American Public Health Association, A.P.H.A. 1995. Standard methods for the examination of water and wastewater (Vol. 21). Washington, DC: American Public Health Association.

Amoah, K., Huang, Q.C., Tan, B.P., Zhang, S., Chi, S.Y., Yang, Q.H., Liu, H.Y. and Dong, X.H. 2019. Dietary supplementation of probiotic *Bacillus coagulans* ATCC 7050, improves the growth performance, intestinal morphology, microflora, immune response, and disease confrontation of Pacific white shrimp, *Litopenaeus vannamei. Fish & Shellfish Immunology*, 87: 796-808. https://doi.org/10.1016/j.fsi.2019.02.029

Bachère, E., Gueguen, Y., Gonzalez, M., De Lorgeril, J., Garnier, J. and Romestand, B. 2004. Insights into the anti microbial defense of marine invertebrates: the Penaeid shrimps and the oyster *Crassostrea gigas. Immunological Reviews*, 198(1): 149-168. https://doi.org/10.1111/j.0105-2896.2004.00115.x

Bernfeld, P. 1955. Amylase α and β. *Methods in Enzymology*, 1: 149-151. https://doi.org/10.1016/0076-6879 (55)01021-5

Buruiana, C.T., Profir, A.G. and Vizireanu, C. 2014. Effects of probiotic *Bacillus* species in aquaculture-an overview. The Annals of the University of Dunarea de Jos of Galati. Fascicle VI. *Food Technology*, 38: p.9.

Castex, M., Lemaire, P., Wabete, N. and Chim, L. 2010. Effect of probiotic *Pediococcus acidilactici* on antioxidant defences and oxidative stress of *Litopenaeus stylirostris* under *Vibrio nigripulchritudo* challenge. *Fish & Shellfish Immunology*, 28(4): 622-631. https://doi.org/10.1016/j.fsi.2009.12.024

Chai, P.C., Song, X.L., Chen, G.F., Xu, H. and Huang, J. 2016. Dietary supplementation of probiotic *Bacillus* PC465 isolated from the gut of *Fenneropenaeus chinensis* improves the health status and resistance of *Litopenaeus vannamei* against White Spot Syndrome virus. *Fish & shellfish immunology*, 54: 602-611. https://doi.org/10.1016/j.fsi.2016.05.011

Cherry, I.S. and Crandall Jr, L.A. 1932. The specificity of pancreatic lipase: its appearance in the blood after pancreatic injury. *American Journal of Physiology-Legacy Content*, 100(2): 266-273. https://doi.org/10.1152/ajplegacy.1932.100.2.266

Chiu, C.H., Guu, Y.K., Liu, C.H., Pan, T.M. and Cheng, W. 2007. Immune responses and gene expression in white shrimp, *Litopenaeus vannamei*, induced by *Lactobacillus plantarum*. *Fish & Shellfish Immunology*, 23(2): 364-377. https://doi.org/10.1016/j.fsi.2006.11.010

Clemente, J.C., Ursell, L.K., Parfrey, L.W. and Knight, R. 2012. The impact of the gut microbiota on human health: an integrative view. *Cell*, 148(6): 1258-1270. https://doi.org/10.1016/j.cell.2012.01.035

Dawood, M.A., Koshio, S. and Esteban, M.Á. 2018. Beneficial roles of feed additives as immunostimulants in aquaculture: a review. *Reviews in Aquaculture*, *10*(4): 950-974. https://doi.org/10.1111/req.12209

Dharmaraj, R. and Rajendren, V. 2014. Probiotic assessment of *Bacillus infantis* isolated from gastrointestinal tract of *Labeo rohita*. *International Journal of Scientific and Research Publications*, 4(7): 1-6.

Drapeau, G.R. 1976. Protease from *Staphyloccusaureus*. In Methods in enzymology, 45: 469-475. Academic Press. https://doi.org/10.1016/S0076-6879(76)45041-3

Ferreira, G.S., Bolívar, N.C., Pereira, S.A., Guertler, C., do Nascimento Vieira, F., Mouriño, J.L.P. and Seiffert, W.Q. 2015. Microbial biofloc as source of probiotic bacteria for the culture of *Litopenaeus vannamei*. *Aquaculture*, 448: 273-279. https://doi.org/10.1016/j.aquaculture.2015.06.006

Food and Agriculture Organization of the United Nations (FAO)/ World Health Organization (WHO). 2001. Joint FAO/WHO expert consultation on evaluation of health and nutritional properties of probiotics in food including powder milk and live lactic acid bacteria.

García Bernal, M., Medina Marrero, R., Rodríguez Jaramillo, C., Marrero Chang, O., Campa Córdova, Á.I., Medina García, R. and Mazón Suástegui, J.M. 2018. Probiotic effect of *Streptomyces* spp. on shrimp (*Litopenaeus vannamei*) post larvae challenged with *Vibrio parahaemolyticus*. *Aquaculture Nutrition*, 24(2): 865-871. https://doi.org/10.1111/anu.12622

Gollas-Galván, T., Hernández-López, J. and Vargas-Albores, F. 1997. Effect of calcium on the prophenoloxidase system activation of the brown shrimp (*Penaeus californiensis*, Holmes). Comparative Biochemistry and Physiology Part A: *Physiology*, 117(3): 419-425. https://doi.org/10.1016/S0300-9629(96)00363-5

González-Peña, M.C., A.J. Anderson, D.M. Smith & G.S. Moreira. 2002. Effect of dietary cellulose on digestion in the prawn *Macrobrachium rosenbergii*. *Aquaculture*, 211: 291-303. https://doi.org/10.1016/S0044-8486 (02)00129-1

Gullian, M., Thompson, F. and Rodriguez, J. 2004. Selection of probiotic bacteria and study of their immunostimulatory effect in *Penaeus vannamei*. *Aquaculture*, 233(1-4): 1-14. https://doi.org/10.1016/j.aquaculture.2003.09.013

Interaminense, J.A., Vogeley, J.L., Gouveia, C.K., Portela, R.W., Oliveira, J.P., Andrade, H.A., Peixoto, S.M., Soares, R.B., Buarque, D.S. and Bezerra, R.S. 2018. In vitro and in vivo potential probiotic activity of *Bacillus subtilis* and *Shewanella* algae for use in *Litopenaeus vannamei* rearing. *Aquaculture*, 488: 114-122. https://doi.org/10.1016/j.aquaculture.2018.01.027

Kar, N. and Ghosh, K. 2008. Enzyme producing bacteria in the gastrointestinal tracts of *Labeo rohita* (Hamilton) and *Channa punctatus* (Bloch). *Turkish Journal of Fisheries and Aquatic Sciences*, 8(1): 115-120.

Khademzade, O., Zakeri, M., Haghi, M. and Mousavi, S.M. 2020. The effects of water additive *Bacillus cereus* and *Pediococcus acidilactici* on water quality, growth performances, economic benefits, immunohematology and bacterial flora of white leg shrimp (*Penaeus vannamei* Boone, 1931) reared in earthen ponds. *Aquaculture Research*, 51(5): 1759-1770. https://doi.org/10.1111/are.14525

Kirchgessner, M., Roth, F.X., Eidelsburger, U. and Gedek, B. 1993. The nutritive efficiency of *Bacillus cereus* as a probiotic in the raising of piglets. 1. Effect on the growth parameters and gastrointestinal environment. *Archiv fur Tierernahrung*, 44(2): 111-121. https://doi.org/10.1080/17450399309386062

Krummenauer, D., Poersch, L., Romano, L.A., Lara, G.R., Encarnação, P. and WasieleskyJr, W. 2014. The effect of probiotics in a *Litopenaeus vannamei* biofloc culture system infected with *Vibrio parahaemolyticus*. *Journal of applied Aquaculture*, 26(4): 370-379. https://doi.org/10.1080/10454438.2014.965575

Lafferty, K.D., Harvell, C.D., Conrad, J.M., Friedman, C.S., Kent, M.L., Kuris, A.M., Powell, E.N., Rondeau, D. and Saksida, S.M. 2015. Infectious diseases affect marine fisheries and aquaculture economics. *Annual Review of Marine Science*, 7: 471-496. https://doi.org/10.1146/annurev-marine-010814-015646

Liu, C.H., Chiu, C.S., Ho, P.L. and Wang, S.W. 2009. Improvement in the growth performance of white shrimp, *Litopenaeus vannamei*, by a protease producing probiotic, *Bacillus subtilis* E20, from natto. *Journal of Applied Microbiology*, 107(3): 1031-1041. https://doi.org/10.1111/j.1365-2672.2009.04284.x

Lowry, O.H., Rosebrough, N.J., Farr, A.L. and Randall, R.J. 1951. Protein measurement with the Folin phenol reagent. *Journal of Biological Chemistry*, 193: 265-275.

Luis-Villasenor, I.E., Macías-Rodríguez, M.E., Gómez-Gil, B., Ascencio-Valle, F. and Campa-Córdova, Á.I. 2011. Beneficial effects of four *Bacillus* strains on the larval cultivation of *Litopenaeus vannamei*. *Aquaculture*, 321(1-2): 136-144. https://doi.org/10.1016/j.aquaculture.2011.08.036

Majeed, M., Nagabhushanam, K., Natarajan, S., Sivakumar, A., Ali, F., Pande, A., Majeed, S. and Karri, S.K. 2015. *Bacillus coagulans* MTCC 5856 supplementation in the management of diarrhoea predominant Irritable Bowel Syndrome: a double blind randomized placebo controlled pilot clinical study. *Nutrition Journal*, 15(1): 21. https://doi.org/10.1186/s12937-016-0140-6

Misra, H.P. and Fridovich, I. 1972. The role of superoxide anion in the autoxidation of epinephrine and a simple assay for superoxide dismutase. *Journal of Biological chemistry*, 247(10): pp.3170-3175. https://www.jbc.org/content/247/10/3170.short

Muñoz, M., Cedeño, R., Rodrýìguez, J., van der Knaap, W.P., Mialhe, E. and Bachère, E. 2000. Measurement of reactive oxygen intermediate production in haemocytes of the penaeid shrimp, *Penaeus vannamei*. *Aquaculture*, 191(1-3): 89-107. https://doi.org/10.1016/S0044-8486 (00)00420-8

Musthafa, K.S., Saroja, V., Pandian, S.K. and Ravi, A.V. 2011. Antipathogenic potential of marine *Bacillus* sp. SS4 on N-acyl-homoserine-lactone-mediated virulence factors production in *Pseudomonas aeruginosa* (PAO1). *Journal of Biosciences*, 36(1): 55-67. https://doi.org/10.1007/s12038-011-9011-7

Nayak, S.K. 2010. Probiotics and immunity: a fish perspective. *Fish & Shellfish Immunology*, 29(1): 2-14. https://doi.org/10.1016/j.fsi.2010.02.017

Nimrat, S., Khaopong, W., Sangsong, J., Boonthai, T. and Vuthiphandchai, V. 2019. Dietary administration of *Bacillus* and yeast probiotics improves the growth, survival, and microbial community of juvenile white leg shrimp, *Litopenaeus vannamei*. *Journal of Applied Aquaculture*, 1-17. https://doi.org/10.1080/10454438.2019.1655517

Nyangale, E.P., Farmer, S., Keller, D., Chernoff, D. and Gibson, G.R. 2014. Effect of prebiotics on the fecal microbiota of elderly volunteers after dietary supplementation of *Bacillus coagulans* GBI-30, 6086. *Anaerobe*, 30: 75-81. https://doi.org/10.1016/j.anaerobe.2014.09.002

Olmos, J., Ochoa, L., Paniagua-Michel, J. and Contreras, R. 2011. Functional feed assessment on *Litopenaeus vannamei* using 100% fish meal replacement by soybean meal, high levels of complex carbohydrates and *Bacillus* probiotic strains. *Marine Drugs*, 9. https://doi.org/10.3390/md9061119

Runsaeng, P., Kwankaew, P. and Utarabhand, P. 2018. FmLC6: An ultimate dual-CRD C-type lectin from *Fenneropenaeus merguiensis* mediated its roles in shrimp defence immunity towards bacteria and virus. *Fish & shellfish immunology*, 80: 200-213. https://doi.org/10.1016/j.fsi.2018.05.043

Schubert, R.H., Buchanan, R.E. and Gibbons, N.E. 1974. Bergey's Manual of Determinative Bacteriology. by RE Buchanan and NE Gibbons, p.354.

Schulz, O., Diebold, S.S., Chen, M., Näslund, T.I., Nolte, M.A., Alexopoulou, L., Azuma, Y.T., Flavell, R.A., Liljeström, P. and e Sousa, C.R. 2005. Toll-like receptor 3 promotes cross-priming to virus-infected cells. *Nature*, 433(7028): 887. https://doi.org/10.1038/nature03326

Selim, K.M. and Reda, R.M. 2015. Improvement of immunity and disease resistance in the Nile tilapia, *Oreochromis niloticus*, by dietary supplementation with *Bacillus amyloliquefaciens*. *Fish & Shellfish Immunology*, 44(2): 496-503. https://doi.org/10.1016/j.fsi.2015.03.004

Shen, W.Y., Fu, L.L., Li, W.F. and Zhu, Y.R. 2010. Effect of dietary supplementation with *Bacillus subtilis* on the growth, performance, immune response and antioxidant activities of the shrimp (*Litopenaeus vannamei*). *Aquaculture Research*, 41(11): 1691-1698. https://doi.org/10.1111/j.1365-2109.2010.02554.x

Subramenium, G.A., Swetha, T.K., Iyer, P.M., Balamurugan, K. and Pandian, S.K. 2018. 5-hydroxymethyl-2-furaldehyde from marine bacterium *Bacillus subtilis* inhibits biofilm and virulence of *Candida albicans*. *Microbiological research*, 207: 19-32. https://doi.org/10.1016/j.micres.2017.11.002

Takahara, S., Hamilton, H.B., Neel, J.V., Kobara, T.Y., Ogura, Y. and Nishimura, E.T. 1960. Hypocatalasemia: a new genetic carrier state. *The Journal of Clinical Investigation*, 39(4): 610-619.

Tavares-Sánchez, O.L., Gómez-Anduro, G.A., Felipe-Ortega, X., Islas-Osuna, M.A., Sotelo-Mundo, R.R., Barillas-Mury, C. and Yepiz-Plascencia, G. 2004. Catalase from the white shrimp Penaeus (*Litopenaeus vannamei*): molecular cloning and protein detection. Comparative Biochemistry and Physiology Part B: *Biochemistry and Molecular Biology*, 138(4): 331-337. https://doi.org/10.1016/j.cbpc.2004.03.005

Thitamadee, S., Prachumwat, A., Srisala, J., Jaroenlak, P., Salachan, P.V., Sritunyalucksana, K., Flegel, T.W. and Itsathitphaisarn, O. 2016. Review of current disease threats for cultivated Penaeid shrimp in Asia. *Aquaculture*, 452: 69-87. https://doi.org/10.1016/j.aquaculture.2015.10.028

Touraki, M., Karamanlidou, G., Karavida, P. and Chrysi, K. 2012. Evaluation of the probiotics *Bacillus subtilis* and *Lactobacillus plantarum* bioencapsulated in *Artemia* nauplii against vibriosis in European sea bass larvae (*Dicentrarchus labrax*). *World Journal of Microbiology and Biotechnology*, 28(6): 2425-2433. https://doi.org/10.1007/s11274-012-1052-z

Tseng, D.Y., Ho, P.L., Huang, S.Y., Cheng, S.C., Shiu, Y.L., Chiu, C.S. and Liu, C.H. 2009. Enhancement of immunity and disease resistance in the white shrimp, *Litopenaeus vannamei*, by the probiotic, *Bacillus subtilis* E20. *Fish & Shellfish Immunology*, 26(2): 339-344. https://doi.org/10.1016/j.fsi.2008.12.003

Veron, H.E., Di Risio, H.D., Isla, M.I. and Torres, S. 2017. Isolation and selection of potential probiotic lactic acid bacteria from *Opuntiaficus-indica* fruits that grow in Northwest Argentina. LWT, 84: 231-240. https://doi.org/10.1016/j.lwt.2017.05.058

Verschuere, L., Rombaut, G., Sorgeloos, P. and Verstraete, W., 2000. Probiotic bacteria as biological control agents in aquaculture. *Microbiol. Microbiology and Molecular Biology Reviews*, 64(4): 655-671. 10.1128/MMBR.64.4.655-671.2000

Vieira, F.D.N., Jatobá, A., Mouriño, J.L.P., BuglioneNeto, C.C., Silva, J.S.D., Seiffert, W.Q., Soares, M. and Vinatea, L.A. (2016). Use of probiotic-supplemented diet on a Pacific white shrimp farm. *Revista Brasileira de Zootecnia*, 45(5): 203-207. https://doi.org/10.1590/S1806-92902016000500001

Wang, Y.B. 2007. Effect of probiotics on growth performance and digestive enzyme activity of the shrimp *Penaeus vannamei*. *Aquaculture*, 269(1-4): 259-264. https://doi.org/10.1016/j.aquaculture.2007.05.035

Wang, Y.C., Hu, S.Y., Chiu, C.S. and Liu, C.H. 2019. Multiple-strain probiotics appear to be more effective in improving the growth performance and health status of white shrimp, *Litopenaeus vannamei*, than single probiotic strains. *Fish & Shellfish Immunology*, 84: 1050-1058. https://doi.org/10.1016/j.fsi.2018.11.017

Yang, C., Chowdhury, M.A., Huo, Y. and Gong, J. 2015. Phytogenic compounds as alternatives to in-feed antibiotics: potentials and challenges in application. *Pathogens*, 4(1): 137-156. https://doi.org/10.3390/pathogens4010137

Yengkokpam, S., Debnath, D., Sahu, N.P., Pal, A.K., Jain, K.K. and Baruah, K. 2016. Dietary protein enhances non-specific immunity, anti-oxidative capability and resistance to *Aeromonas hydrophila* in *Labeo rohita* fingerlings pre-exposed to short feed deprivation stress. *Fish & shellfish immunology*, 59: 439-446. https://doi.org/10.1016/j.fsi.2016.10.052

Ziaei-Nejad, S., Rezaei, M.H., Takami, G.A., Lovett, D.L., Mirvaghefi, A.R. and Shakouri, M. 2006. The effect of *Bacillus* spp. bacteria used as probiotics on digestive enzyme activity, survival and growth in the Indian white shrimp *Fenneropenaeus indicus*. *Aquaculture*, 252(2-4): 516-524. https://doi.org/10.1016/j.aquaculture.2005.07.021

Zokaeifar, H., Babaei, N., Saad, C.R., Kamarudin, M.S., Sijam, K. and Balcazar, J.L. 2014. Administration of *Bacillus subtilis* strains in the rearing water enhances the water quality, growth performance, immune response, and resistance against *Vibrio harveyi* infection in juvenile white shrimp, *Litopenaeus vannamei*. *Fish & Shellfish Immunology*, 36(1): 68-74. https://doi.org/10.1016/j.fsi.2013.10.007

Zokaeifar, H., Balcázar, J.L., Saad, C.R., Kamarudin, M.S., Sijam, K., Arshad, A. and Nejat, N. 2012. Effects of *Bacillus subtilis* on the growth performance, digestive enzymes, immune gene expression and disease resistance of white shrimp, *Litopenaeus vannamei*. *Fish & Shellfish Immunology*, 33(4): 683-689. https://doi.org/10.1016/j.fsi.2012.05.027

CHAPTER 7

EXTRA CELLULAR POLYMER SUBSTANCES (EPS) PRODUCING BACTERIA FROM BIOFLOC OF NILE TILAPIA CULTURE SYSTEM USING DISTILLERY SPENT WASH AS CARBON SOURCE

The present study aimed to isolate and characterize the Extracellular polymeric substance (EPS) producing bacteria from biofloc reared Nile Tilapia ponds. Distillery spent wash was used as a carbon source to maintain the C:N ratio at 10:1 in the fish culture ponds and screening of bacteria were done fortnightly in 180 days culture. Out of 38 bacterial isolates, 7 isolates were found to produce EPS. Based on 16s rRNA sequence analysis the isolates were identified as *Bacillus subtilis, B. megaterium, B. infantis, B.cereus, Pseudomonas balearica, P.mendocina* and *P. alcaligenes.* The highest production of EPS was recorded in *B.cereus* (1.25g/L). EPS extracted from Bacillus cereus was reported to have higher protein (89μg/ml) and *B.*

subtilis possessed higher carbohydrate(753.75µg/ml). Maximum flocculating ability of 40.18% in *B. cereus* and higher emulsifying activity of 63.53% was observed in *B. megaterium*. The EPS extracted from *B.infantis* showed lower sludge volume index on its treatment with aquaculture sludge (15.38 ml/g). Absorption band in the range of 4000/cm to 450/cm using FTIR analysis confirmed the presence of characteristic functional bands arising from polysaccharides, nucleic acids and proteins. The results indicated the presence of EPS producing bacteria in biofloc based Nile Tilapia aquaculture systems.

Tilapia is one of the widely cultured fish globally in various aquaculture systems (FAO, 2017). However, tilapia production was hugely affected by emerging disease outbreaks thereby hindering the development of aquaculture industry (Telli *et al.* 2014; Menaga *et al.* 2019). Also, to ensure long-term sustainability, environmental impacts must be minimized and alternative ways such as flocculation process need to be applied. Flocculation helps to overcome the problem of aquaculture effluent by the addition of synthetic polymers. As sludge from aquaculture ponds is negatively charged, they can interact with these synthetic polymers in combination with cations to neutralize the surface charges thereby aiding the flocculation and settling (Higgins and Novack, 1997b). However, these synthetic polymers not only enhance the flocculation and sludge dewatering but also when discharged into the environment affects the soil microorganism as a major disadvantage. To overcome these hindrances, green technology metabolites known as bio flocculants produced by microorganisms can act similar function as synthetic flocculants (Zaki *et al.* 2011). Many microorganisms isolated from sludge were reported to secrete extracellular polymeric substances. They are mainly composed of functional proteins, polysaccharides, protein, nucleic acid and cellulose (Kumar *et al.* 2004; Feng and Xu, 2008). The bacterial extracellular polymeric substances also improve the formation of bioflocs in activated sludge and contribute to its structural, surface charge and settling properties (Houghton *et al.* 2001). Consequently, better understanding on the characteristics of EPS in biofloc provides advantages in undergoing wastewater treatment process. Biofloc technology, a zero-water exchange system, can be used to overcome the problems such as sludge settling by stocking at a higher density with the enhanced fish production (Hargreaves, 2013; Green *et al.* 2019). Therefore, the ultimate aim of this study was to characterize the potential flocculant-producing bacteria particularly for aquaculture wastewater treatment. In the present study 7 bacterial species isolated from Biofloc culture water were identified as EPS producing bacteria. The EPS extracted from the bacteria has been studied for various properties for its application in flocculation and sludge settling.

METHODOLOGY

Isolation and Characterization of EPS Producing Bacteria from Biofloc:

Nile tilapia juveniles (5.5 ±0.21g) were stocked at a density of $10/m^3$ in $300m^2$ ponds for 180 days in triplicates. Biofloc development and maintenance in the freshwater culture ponds were adopted as suggested by Taw(2006). Distillery spentwash obtained from M/s. Rajshree Biosolutions, Coimbatore has been used as a carbon source to maintain the C:N ratio at 10:1. The biofloc water sample from Nile tilapia cultured ponds was collected, serially diluted and spread plated on nutrient agar. The plates were incubated at 37°C. Morphologically different colonies were picked and stained using crystal violet and 20% aqueous $CuSO_4$ solution for initial screening of EPS producing bacteria (Cain *et al.*2009). Morphological characterization was carried out for the selected bacterial isolates according to Bergey's manual (Brown, 1939).

Genotypic Identification and Phylogenetic Tree Construction of EPS Producing Bacteria

The genomic DNA was isolated using Phenol-Chloroform method from EPS producing bacteria. The isolated DNA was amplified for its 16s rRNA region using the Forward primer-AGAGTTTGATCCTGGCTCAG and Reverse primer-CGGTTACCTTGTTACGACTT.The amplified DNA was sequenced and subjected to BLAST analysis. The aligned sequences were submitted in Genbank database to obtain the accession number. The phylogenetic tree was constructed to find the relationship between the EPS producing bacterial isolates.

Extraction of EPS from Bacterial Isolates

The bacterial isolates were incubated in liquid medium specific for EPS production containing 0.2 g KH_2PO_4, 0.8 g K_2HPO_4, 0.2 g $MgSO_4.H_2O$, 0.1g $CaSO_4.2H_2O$, 2 mg $FeCl_3$, $Na_2MoO_4.2H_2O$ (trace), 0.5g Yeast extract, 20 g Sucrose per litre and after 48hrs of incubation the cell pellet were harvested by centrifugation at 5000 rpm for 15 minutes. The cell pellet is resuspended by adding 500 µl of 1mM EDTA and centrifuged at 9000 rpm for 10 minutes. To the cell free supernatant, cold acetone solution (1:3) was added and centrifuged to obtain the pellet. The final pellet was dried at 60°C for 24hrs to obtain the capsular EPS (Mu'minah and Subair, 2015).

Characterization of EPS

Protein and carbohydrate estimation: EPS solution were prepared by harvesting the cell pellets from the specific medium. The pellets were resuspended in twice the volume of 0.2M cold sulphuric acid solution followed by stirring at 4°C for 3 h. The resulting suspension was centrifuged at 15,000 rpm for 20 min to obtain the supernatant (EPS solution) which can be used for further analysis. The amount of protein and carbohydrate in EPS solution was quantified using Lowry's method (1951) and phenol sulphuric method (Dubois *et al.* 1956).

Lipid Emulsifying Test

Emulsifying activity of EPS producing bacterial isolates were tested using the method of Yasumatsu *et al.* (1972). An equal volume of olive oil and EPS solution were incubated on a rotary shaker for 11 days. The lipid emulsifying activity was expressed as percentage of the height of emulsified layer to the height of whole layer.

Flocculation Ability

To 2 ml of EPS solution, 1 ml of 6.8 mM $CaCl_2$ and 10 ml of 5g/l activated charcoal were added and mixed well. Control was prepared without adding EPS solution. The mixture is kept incubated at room temperature and the absorbance was measured at 550 nm. Flocculation activity was calculated according to following formula (Aziz *et al.* 2012).

$$\text{Flocculating activity (\%)} = A\text{-}B/A \times 100$$

Where A and B are the optical density values of control and samples respectively.

Sludge Volume Index

Sludge volume index (SVI) of the EPS producing bacterial isolates was estimated according to the method of Subramanian *et al.* (2010). The sludge sample was collected and 2% of extracted EPS (v/v) was added to this and mixed well. Sludge sample without EPS was considered as control. The settled sludge volume and suspended solids was measured after 30 min.

SVI was Calculated using the Formula

Sludge volume index (SVI) (ml/g) = Settled sludge volume / Suspended solids

Fourier Transform Infra-red Analysis (FTIR): The EPS extracted from bacterial isolates was subjected to Fourier Transform Infra-red Analysis and spectra were recorded in 4000/cm to 450/cm range for confirmation of the functional groups in EPS arising from polysaccharides, nucleic acids and proteins.

RESULTS AND CONCLUSIONS

Isolation and biochemical characterization of EPS producing bacteria: Out of 38 morphologically different colonies only 7 were confirmed for EPS production using crystal violet and 20% aqueous $CuSO_4$ solution stained for initial screening of EPS producing bacteria. The morphological and biochemical characterization of 7 bacterial isolates were listed in the Table 1.

Phylogenetic tree for EPS producing bacteria: The accession numbers obtained from genbank database are *Bacillus infantis* (577 bp) -MH424755, *Bacillus subtilis* (568 bp) -MH424900, *Bacillus megaterium* (577bp) - MH424904,*Pseudomonas alcaligenes* (518 bp) - MH424895, *Bacillus cereus* (492 bp) - MH997476, *Pseudomonas balearica* (736 bp) - MH997474, *Pseudomonas mendocina* (669 bp) - MH997472. It is inferred from the phylogenetic tree that there exist a close relationship between the four bacillus species: *Bacillus infantis, Bacillus cereus, Bacillus megaterium, Bacillus subtilis* and three *Pseudomonas species*: *Pseudomonas mendocina, Pseudomonas alcaligenes, Pseudomonas balearica.*Bacteria in general tend to produce EPS under unfavourable environmental conditions to prevent them from desiccation, toxic compounds, low temperatures or high osmotic pressures and during oxygen and nitrogen level fluctuations (Hirst *et al.* 2003). The EPS obtained in our study was Capsular EPS and its tightly bound with the cell wall by non-covalent linkages that helps in the biofloc or biofilm formation by interacting with the negatively charged sludge solids(Higgins and Novack, 1997b).The results from our study reveals the fact that both gram positive and gram negative bacteria can produce EPS as

Table 1: Morphological and biochemical characterization of bacterial isolates

Biochemical characteristics	*Pseudomonas mendocina*	*P. balearica*	*P. alcaligenes*	*Bacillus cereus*	*B. subtilis*	*B. megaterium*	*B. infantis*
Colony Morphology	-ve rods	-ve rods	-ve rods	+ve rods	+ve rods	+ve rods	+ve rods
Indole	-ve	-ve	-ve	-ve	-ve	-ve	-ve
Methyl red	-ve	-ve	+ve	+ve	-ve	-ve	-ve
VogesProskauer's	-ve	-ve	-ve	+ve	+ve	-ve	+ve
Citrate utilization	+ve	-ve	-ve	-ve	+ve	+ve	+ve
Glucose	+ve	+ve	+ve	+ve	+ve	+ve	+ve
Adonitol	+ve	+ve	+ve	+ve	+ve	-ve	-ve
Arabinose	+ve	+ve	+ve	+ve	+ve	-ve	-ve
Lactose	+ve	+ve	+ve	+ve	+ve	-ve	+ve
Sorbitol	+ve	+ve	+ve	+ve	+ve	-ve	-ve
Mannitol	+ve	+ve	+ve	+ve	+ve	-ve	+ve
Sucrose	+ve	+ve	+ve	+ve	+ve	-ve	+ve

most of the microorganisms in the environment have the ability to live within these biofilms. Biofloc helps the bacteria to form the microbial aggregates as an essential step for the survival (Flemming and Wingender, 2001b) in a huge bacterial population. The anionic nature of EPS can also be augmented by the presence of uronic acids or ketal-linked pyruvates thereby enhancing the interactions of divalent cations such as Calcium and Magnesium which helps in increasing the binding force in the biofilm (Donlan, 2002; Sutherland, 2001).The identified isolates were dominantly *Bacillus sp.* which agrees with the findings of Hashim *et al.* (2018). Previous studies have also reported on the presence of bioflocculating bacteria in biofloc culture ponds of *P. vannamei* (Kasan *et al.* 2016).

Extraction of EPS from bacterial isolates: The EPS extracted from seven bacterial isolates were listed in Table 2. The EPS extracted from bacterial isolates of biofloc culture water varies from 0.33-1.25g/L with the maximum production in *Bacillus cereus*(1.25g/L).The maximum EPS production (3.2g/L) from *Bacillus cereus* isolated from sludge water (Subramaniam *et al.* 2007) was also previously reported. The variable production of EPS by different isolates was may be due to the bacterial metabolic activity (Subramanian *et al.* 2010), genetic organization of gene clusters, biosynthesis of sugar precursors and regulatory elements (Nouha *et al.* 2016).

Carbohydrate and protein content in EPS: The carbohydrate and protein content of EPS producing bacterial isolates were listed in Table 2. The highest carbohydrate was found to be present in EPS extracted from *B. subtilis* and the protein was higher in the EPS of *B. cereus.* Total carbohydrate concentration was higher than the total protein in all the extracted EPS from the bacterial isolates which is in agreement with the previous investigations of Shahnavaz *et al.* (2015) who also showed the dominance of carbohydrate in EPS than protein. This increased concentration of carbohydrate than protein plays a major role in sludge settling by forming the bridges between the negatively charged groups and divalent cations present in the sludge(Higgins and Novack, 1997b).However, proteins also help in the floc formation (Urbain *et al.* 1993) contributing to the binding strength and stability, by hydrophobic interactions and polyvalent cation bridging(Jorand *et al.* 1998).

Table 2. EPS extracted, protein ,carbohydrate content and sludge volume index

Isolates	EPS in grams/L	Protein (µg/ml)	Carbohydrate (µg/ml)	Sludge volume index ml/g
Bacillus subtilis	0.92	40	753.75	27.78
B. megaterium	0.98	26	303.80	21.43
B. infantis	0.33	19	188.75	15.38
B. cereus	1.25	89	377.50	20.10
Pseudomonas balearica	0.6	59	476.25	16.67
P. mendocina	0.483	18	667.50	30.77
P. alcaligenes	0.53	29	71.25	23.08

Flocculation and Emulsifying activity of EPS: The flocculation ability and emulsifying activity of seven EPS producing bacterial isolates were analysed. The maximum flocculation percentage observed in *Bacillus cereus* (40.18%) and this may be de due to the fact that EPS with long polymeric chains holds more active sites for binding when compared with the EPS produced by the other bacterial isolates. Also the maximum flocculating ability can be related to the highest production of EPS by the *Bacillus cereus.* The higher flocculating ability has been previously reported in *Bacillus cereus* isolated from biofloc culture ponds of Pacific white leg shrimp (93%) by Ismail and Amin (2018). *Bacillus megaterium* was found to have the highest emulsifying activity (63.53%) which was probably due to the higher levels of uric acid content in the extracted EPS of this specific isolate. This is in agreement with the studies of Yun and Park (2003) as they reported the polysaccharide produced by *Bacillus sp.* CP912 has a greater potential to use as an emulsifier. Results from the previous findings correlated a significant increase in the emulsifying activity of EPS showed higher biodegradability and lower solubility in water. Thus the glycol protein nature of the EPS can be used as excellent emulsifying agents with various medical and environmental applications (Cameotra and Makkar, 2004; Rosenberg and Ron, 1999).

Sludge volume index: The Sludge volume index of EPS producing bacteria were listed in table 2. The EPS of *B.infantis* showed lower sludge volume index when it is treated with sludge water collected from Nile Tilapia biofloc culture ponds (15.38 ml/g).Sludge Volume Index was found to be below 150mL/g for all the bacterial isolates which is required for a good sludge settling (APHA, 2005).Similarly Subramanian *et al.* (2010) isolated *Pseudomonas sp* from waste water sludge and proved the ability of these bacterial strains in good sludge settling which is in accordance with our study as *Pseudomonas balearica* has the second lowest SVI followed by *Bacillus infantis.*

Fourier Transform Infra-red Analysis (FTIR): FTIR spectra of the extracted EPS from bacterial isolates were obtained. The peaks for *Bacillus subtilis* 3432.05/cm (Fig.1) , *B.magaterium* 3449.20/cm (Fig. 2), *B. infantis* 3438.42/cm, *B.cereus* 3450.06/ cm, *Pseudomonas balearica* 3453.52/cm, *P. mendocina* 3453.25 / cm, and *P. alcaligenes* 3417.87/cm were obtained indicating hydroxyl group from carbohydrate and water molecules present in the EPS(Liang *et al.* 2010). The peaks for alkyl groups were recorded at 2924.14/cm for *Bacillus subtilis,* 2924.45/cm for *B. megaterium,* 2922.68/cm for *B. infantis,* 2925.05/cm for *B. cereus,* 2924.35/cm for *Pseudomonas balearica,* 2924.08/cm for *P mendocina,* and 2924.58/cm for *P. alcaligenes* (Brian- Jaisson *et al.* 2016).The peaks for carboxylic groups were recorded at 1647.67/cm for *Bacillus subtilis,* 1647.76/cm for *B. megaterium,* 1644.45/cm for *B. infantis,* 1647.86/cm for *B. cereus,* 1647.35/cm for *Pseudomonas balearica,* 1644.95/cm for *P mendocina,* and 1645.20/cm for *P. alcaligenes* (Caruso *et al.,* 2018).FTIR peaks also proved the characteristic bands arising from the functional groups present in the EPS which was also previously confirmed from the studies of Brian - Jaisson *et al.* (2016). In the present study EPS producing isolates were screened from the culture ponds of Nile Tilapia. The various characteristics of the EPS secreted by the isolates proved that it not only helps in the floc and filament formation in biofloc culture ponds but also in sludge settling as a mode of aquaculture waste water treatment.

The current study proved the presence of EPS producing bacteria in the biofloc ponds providing the optimal environment for the flocculation and sludge settlement. The identified isolates can be mass cultured and deployed in the bioremediation of

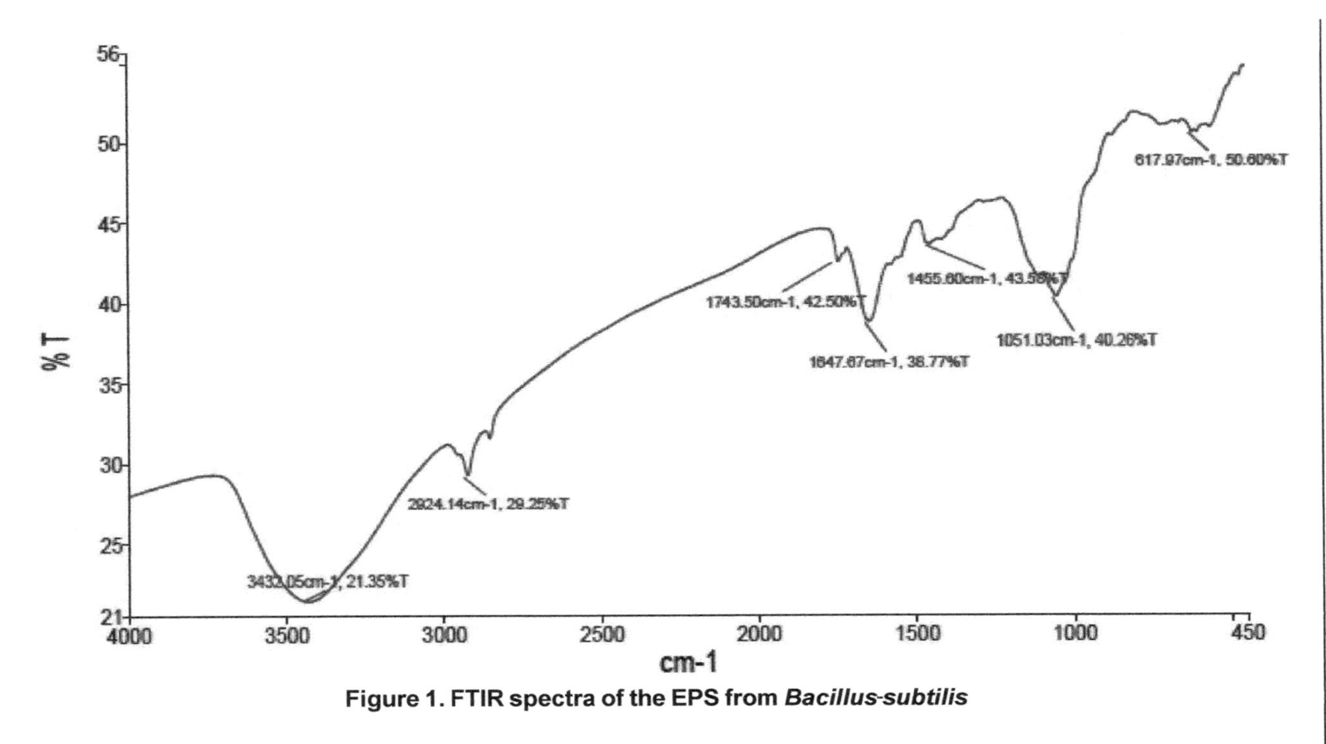

Figure 1. FTIR spectra of the EPS from *Bacillus-subtilis*

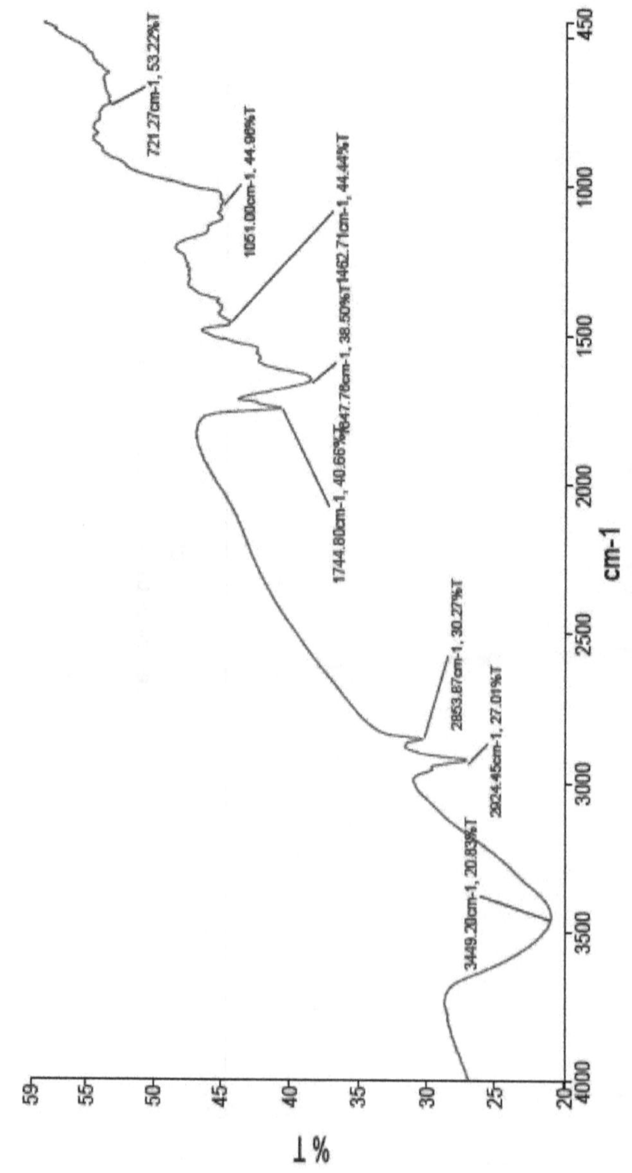

Figure 2. FTIR spectra of the EPS from *Bacillus megaterium*

aquaculture effluents. This study also elucidates the advantage of adopting biofloc technology in aquaculture and more indepth insights on the mechanism in the EPS formation and spatial distribution will help for the broad range of applications in aquaculture.

ACKNOWLEDGEMENT

This research work was part of the Project funded by the Department of Biotechnology, Govt of India (Project code: DBT-507; PI: S.Felix).

REFERENCES

APHA. 2005.*Standard Methods for the Examination of Water and Wastewater.* 21st ed., Washington, DC.

Aziz, A., Shadia, M., Hamed, H.A. and Mouafi, F.E.. 2012. Acidic exopolysaccharideflocculant produced by the fungus *Mucorrouxii*using beet-molasses.*Research in Biotechnology* 3(6): 1–13.

Brian-Jaisson, F., Molmeret, M., Fahs, A., Guentas-Dombrowsky, L., Culioli, G., Blache, Y., Cérantola, S. and Ortalo-Magné, A. 2016. Characterization and anti-biofilm activity of extracellular polymeric substances produced by the marine biofilm-forming bacterium *Pseudoalteromonas ulvae* strain TC14. *Biofouling* 32(5): 547–60.

Brown, J.H. 1939. Bergey's manual of determinative bacteriology.

Cain, D., Hanks, H., Weis, M., Bottoms, C. and Lawson, J. 2009. Microbiology laboratory manual. *Collin County Community College District, McKinney, TX.*

Cameotra, S.S, and Makkar, R.S. 2004. Recent applications of biosurfactants as biological and immunological molecules. *Current opinion in microbiology* 7(3): 262–66.

Caruso, C., Rizzo, C., Mangano, S., Poli, A., Di Donato, P., Finore, I., Nicolaus, B., Di Marco, G., Michaud, L. and Giudice, A.L. 2018. Production and biotechnological potential of extracellular polymeric substances from sponge-associated Antarctic bacteria. *Appl. Environ. Microbiol* 84(4): 1624–17.

Donlan, R.M. 2002. Biofilms: microbial life on surfaces. *Emerging infectious diseases*8(9): 881.

Dubois, M., Gilles, K.A., Hamilton, J.K., Rebers, P.T. and Smith, F. 1956. Colorimetric method for determination of sugars and related substances. *Analytical chemistry*28(3): 350–56.

Feng, D.L. and Xu, S.H. 2008. Characterization of bioflocculant MBF3-3 produced by an isolated Bacillus sp. *World Journal of Microbiology and Biotechnology* 24(9): 1627.

Flemming, H.C. and Wingender, J. 2001. Relevance of microbial extracellular polymeric substances (EPSs)-Part II: Technical aspects. *Water Science and Technology* 43(6): 9–16.

Food and Agriculture Organization of the United Nations (FAO). 2017. Cultured Aquatic Species Information Programme *Oreochromis niloticus* (Linnaeus, 1758).

Green, B.W., Rawles, S.D., Schrader, K.K., Gaylord, T.G. and McEntire, M.E. 2019. Effects of dietary protein content on hybrid tilapia (*Oreochromis aureus* × *O.niloticus*) performance, common microbial off-flavor compounds, and water quality dynamics in an outdoor biofloc technology production system. *Aquaculture* 503: 571–82.

Hargreaves, J.A. 2013. Biofloc production systems for aquaculture.

Higgins, M.J. and Novak, J.T. 1997. Characterization of exocellular protein and its role in bioflocculation. *Journal of Environmental Engineering* 123(5): 479–85.

Hirst, C.N., Cyr, H. and Jordan, I.A. 2003. Distribution of exopolymeric substances in the littoral sediments of an oligotrophic lake. *Microbial Ecology* 46(1): 22-32.

Houghton, J.J., Quarmby, J. and Stephenson, T. 2001. Municipal wastewater sludge dewaterability and the presence of microbial extracellular polymer. *Water Science and Technology* 44(2-3): 373–79.

Hashim, N.F.C., Ghazali, N.A., Amin, N.M., Ismail, N. and Kasan, N.A. 2018. Description of novel marine bioflocculant-producing bacteria isolated from biofloc of Pacific whiteleg shrimp, Litopenaeus vannamei culture ponds. *Bio. Rxiv.* 402065.

Jorand, F., Boue-Bigne, F., Block, J.C. and Urbain, V. 1998. Hydrophobic/hydrophilic properties of activated sludge exopolymeric substances. *Water Science and Technology* 37(4-5): 307–15.

Kasan, N.A., Teh, C., Amin, M.F., Ghazali, N.A., Hashim, N.F.C., Ibrahim, Z. and Amin, N.M., 2016. Isolation of bioflocculant-producing bacteria from *Penaeus vannamei* ponds for the production of extracellular polymeric substances. *Aquaculture, Aquarium, Conservation & Legislation-International Journal of the Bioflux Society (AACL Bioflux)* 9(6).

Kumar, C.G., Joo, H.S., Choi, J.W., Koo, Y.M. and Chang, C.S. 2004. Purification and characterization of an extracellular polysaccharide from haloalkalophilic Bacillus sp. I-450. *Enzyme and Microbial Technology* 34(7): 673–81.

Liang, Z., Li, W., Yang, S. and Du, P. 2010. Extraction and structural characteristics of extracellular polymeric substances (EPS), pellets in autotrophic nitrifying biofilm and activated sludge. *Chemosphere* 81(5): 626–32.

Lowry, O.H., Rosebrough, N.J., Farr, A.L. and Randall, R.J. 1951. Measurement of protein with the Folin phenol reagent. *Journal of Biological Chemistry* 193(1): 265–75.

Menaga, M., Felix, S. and Charulatha, M. 2019. *In vitro* probiotic properties of Bacillus sp isolated from biofloc reared genetically improved farmed tilapia (*Oreochromis niloticus*). *Indian Journal of Animal Sciences* 89(5): 588–93.

Mu'minah, B. and Subair, H. 2015. Isolation and screening bacterial exopolysaccharide (EPS) from potato rhizosphere in highland and the potential as a Producer Indole Acetic Acid (IAA). *Procedia Food Science* 3: 74–81.

Nouha, K., R.D.T and R.Y.S. 2016. EPS producing microorganisms from municipal wastewater activated sludge.

Rosenberg, E. and Ron, E.Z. 1999. High-and low-molecular-mass microbial surfactants. *Applied Microbiology and Biotechnology* 52(2): 154–62.

Shahnavaz, B., Karrabi, M., Maroof, S. and Mashreghi, M. 2015. Characterization and molecular identification of extracellular polymeric substance (EPS) producing bacteria from activated sludge. *Journal of Cell and Molecular Research* 7(2): 86–93.

Subramaniam, S.B., Yan, S., Tyagi, R.D. and Surampalli, R.Y. 2007. Characterization of extracellular polymeric substances (EPS) extracted from both sludge and pure bacterial strains isolated from wastewater sludge for sludge dewatering. *Water Research* 12: 1–7.

Subramanian, S.B., Yan, S., Tyagi, R.D. and Surampalli, R.Y. 2010. Extracellular polymeric substances (EPS) producing bacterial strains of municipal wastewater sludge: isolation, molecular identification, EPS characterization and performance for sludge settling and dewatering. *Water Research* 44(7): 2253–66.

Sutherland, I.W. 2001. Biofilm exopolysaccharides: a strong and sticky framework. *Microbiology*147(1): 3–9.

Taw, N. 2006. Shrimp production in ASP system, CP Indonesia: Development of the technology from R&D to commercial production. *Aquaculture America.*

Telli, G.S., Ranzani-Paiva, M.J.T., Dias, D.C., Sussel, F.R., Ishikawa, C.M. and Tachibana, L. 2014. Dietary administration of *Bacillus subtilis* on hematology and non-specific immunity of Nile tilapia *Oreochromis niloticus* raised at different stocking densities. *Fish & Shellfish Immunology* 39: 305–11.

Urbain, V., Block, J.C. and Manem, J. 1993. Bioflocculation in activated sludge: an analytic approach. *Water Research* 27(5): 829–38.

Yasumatsu, K., Sawada, K., Moritaka, S., Misaki, M., Toda, J., Wada, T. and Ishii, K. 1972. Whipping and emulsifying properties of soybean products. *Agricultural and Biological Chemistry* 36(5): 719–27.

Yun, U.J. and Park, H.D. 2003. Physical properties of an extracellular polysaccharide produced by Bacillus sp. CP912. *Letters in Applied Microbiology* 36(5): 282–87.

Zaki, S., Farag, S., Elreesh, G.A., Elkady, M., Nosier, M. and El Abd, D. 2011. Characterization of bioflocculants produced by bacteria isolated from crude petroleum oil. *International Journal of Environmental Science and Technology* 8(4): 831–40.

CHAPTER 8

EX-SITU PRODUCTION OF BIOFLOC IN POND RACEWAYS AND PHOTOBIOREACTOR

Ecologically sound management techniques are getting popular for the sustainable production of the aquatic animals to meet the raising global food demand. Biofloc based farming is one of the sustainable techniques where manipulation of Carbon Nitrogen ratio in the culture system through external application of carbon sources or lowering the protein level in the feed stimulate the floc production process. Major deal of research demonstrates that *in-situ* based biofloc culture enhances the growth rate (Wasielesky *et al.*, 2006; Crab *et al.*, 2012) decreases food conversion ratio and additional cost associated with it (Burford *et al.*, 2004), digestive enzymes (Xu and Pan *et al.*, 2013) and positive immunological responses (Kim *et al.*, 2014; Kumar *et al.*, 2014).However, these *in-situ* based techniques need additional oxygen demands for microbial and algal respiration, in addition to the oxygen demand of culture animal (Tacon *et al.*, 2002; Burford *et al.*, 2004). This added oxygen demand requires additional aerators, which increases the aeration costs compared to the regular aquaculture systems. Further, consumption pattern and feeding efficacy of the biofloc depend on the grazing efficacy of cultured species. Scientific community recently started exploiting biofloc as a feed ingredient attributed to its high nutritional profile. Moreover, in the present scenario where fish meal, one of the major ingredients in aquaculture feed becoming more expensive as capture fisheries is currently overexploited or stagnated, urges the industry to investigate the alternative sources of protein for aquaculture feeds.

Therefore, attempts on the partial or complete replacement of fish meal with microbial based biofloc, a source of rich quality protein and nutrients apart from minerals and other antioxidants can form an excellent aquaculture feed ingredient. Dry biofloc biomass can be harvested after treating nutrient and organic rich farm effluents through C: N manipulation technique. Kuhn *et al.*(2008) generated biofloc from fish farm effluent using a sequencing batch reactor with carbon supplementation and with a membrane biological reactor without carbon supplementation. The quality and quantity of biofloc can be optimized by changing the effluent used, bioreactor methodology and/or carbon source and level. These systems have multiple advantages apart from producing dry biofloc biomass as a potential source of protein in fish or shrimp diets, recycling of effluents reduced the pressure that effluent discharge on coastal water bodies, this alternative feed offer the aquaculture industry a viable option to replace costly fish meal and conventional feed ingredients by cost effective biofloc produced from aquaculture waste materials.

The overall dynamic of BFT results from ecological relationships (commensalism, competition and predation among others) that represent a trophic micro network comprised of rotifers, ciliates, heterotrophic bacteria and micro algae (Collazos Lasso & Arisa-Castellaos;2015).The latter two groups tend to be the most abundant within the biofloc community.

Biofloc production is largely being undertaken in culture systems itself in an *in-situ* mode till now. For the first time an attempt was made using appropriately designed production systems to develop bio-floc through *in-situ* mode. They are,

a. In-pond Raceways
b. Photobioreactors
c. Sequence batch Reactors

The authors seek to present the state-of-the-art theory concerning the various aspects of the activated sludge system and to develop procedures for optimized cost-based design and operation. An effort to treat the aquaculture effluent was made by using in-pond raceways and Photobioreactor in the Advanced Research Farm Facility, Madhavaram of Tamil Nadu Fisheries University. The system used for treatment of effluents and its efficiency depends on a number of factors including,

i. Bioreactor or Effluent Treatment Plant (ETP) design
ii. Seasonal or year-round ammoniacal discharges and its operation in reactors
iii. Range of temperatures
iv. Desired aquaculture effluent concentration for ammonium ions

v. Other effluents, water quality requirements

vi. Costs

The predominance of the activated sludge systems for the production of biofloc meal has been consolidated, as cost efficient and reliable biological removal of suspended solids, organic material and macro nutrients *viz.* nitrogen and phosphorus has been demonstrated using in-pond raceways and photobioreactor. A steady state model is developed which will be extremely useful for the optimization of the activated sludge from the aquaculture effluents. This model describes the removal of organic material in the activated sludge system and its consequences for the principal parameters determining process performance on effluent quality, excess sludge production and oxygen consumption.

a. In-pond Raceways

The in-pond raceways are used for the biofloc production by pumping in the culture water or 'the used water' from the aquaculture systems. In this design, the aeration intensity is maintained sufficiently to avoid sludge settling to ensure the maintenance of uniform sludge suspension. They are distinguished from other activated sludge variants by the fact that they do not have a final settler or another mechanism to retain the activated sludge. Therefore, in the in-pond raceways the sludge age is always equal to the liquid retention time. Although the absence of the final settler is an operational and cost effective, the price in terms of effluent quality is high.

The in-pond raceways are larger compared to the photobioreactors in treating the same organic load. The cost per unit volume of in-pond raceways is lower because raceway structure is built with an excavation along with the rudimentary protection against erosion, so that the total cost can be kept smaller. An advantage of the larger volume is that occasional toxic loads may be diluted and hence their effect will be reduced. Similarly, sudden organic and hydraulic overloads can be accommodated more easily. The relative disadvantage of in-pond raceways is that in the absence of a final settler the effluent in principle has the same composition as the mixed effluent so that biodegradable material and suspended solids will be discharged. As a consequence, the effluent quality of in-pond raceways is poor in terms of BOD, COD and TSS concentration.

For the *ex-situ* production of biofloc two raceways of 45 x 9 ft were designed and fabricated using galvanized iron pipes. HDPE liner material of 500gsm was laid over frames. Shaded raceway pond of 45 x 9 x 1.5ft of 17tonnes capacity can be used for the production of biofloc meal to serve as an ingredient for shrimp/fish feed. A total of 10 side frames has been fixed with a GI pipe of ¾ inch and 1 ½ inch (Fig. 1A & 1B). The raceway fabrication arch work is constructed with ½ inch GI pipes and the distance between the pipes were maintained as 4.5 m. The total raceway capacity is 30 tonnes and welding works for the arch set up has been done to cover the entire raceways. The roofing arch of the raceway tank is made and covered by the UV protected HDPE liners to prevent the sunlight intrusion into the raceway. A central partition known as baffle was provided to ensure the circular flow of water (Fig. 2A & 2B). This lined pond is usually shallow (0.25-0.4 m deep) because optical absorption and self-shading by the algal and bacterial cells limits light penetration. Usually, a relatively low cell density is achieved using the raceway pond system (<1 g dry weight/L). The raceway ponds are equipped with one hp paddle wheel aerator of two numbers for the free circulation and the suspension of biofloc. The flow of the culture water can be adjusted based on the process flow of the biofloc production and harvest.

(A) (B)

Figure 1A & 1B. Raceway Tank Fabrication Works and Side Framing

(A) (B)

Figure 2A & 2B. A Inside view of the Biofloc production Raceway Ponds

A detachable harvest chamber has been designed to harvest the biofloc produced in the raceways. The inlet of the harvest chamber is connected with a outlet of raceways and collection of biofloc meal is facilitated with the submersible pump. Excess water drained out from the harvest chamber is taken back to the raceways for further fertilization . The system is characterized by the highly turbulent flow, thin layer of suspension (less than 1 cm), and high ratio of exposed surface area to total volume, and thus can achieve dramatically higher volumetric yield (up to 40 g/L) than open ponds. However, the overall yield obtained from this system is around 20–25 g/m^2/day.

Protocol for Biofloc Production in In-pond raceways

- Used water from fish/shrimp ponds after proper screening of toxic compounds and waste solids has to be pumped in the raceway ponds
- Addition of inorganic fertilizers for the enrichment of culture water to produce biofloc has to be undertaken
- Optimal C:N ratio of 20:1 is recommended for large scale biofloc production
- Regular addition of carbon sources for the maintenance of C:N ratio has to be monitored by taking TAN concentration
- Floc volume in the raceway ponds can be assessed by using the Imhoff cone based on the settlement volume of floc in 1000 ml.
- After 10 days notable increase in the floc volume of more than 30 ml per liter can be observed in ponds
- Harvesting of floc can be initiated once the floc volume reaches above 50 ml/L.

- The culture water has to be pumped to the harvesting chamber fitted with nylon mesh of different mesh size.
- On an average,5-8 kg of floc in wet weight can be harvested per day from a 17 tonnes capacity raceway ponds.
- Generally, 50 micron mesh is recommended for harvesting the biofloc from raceway ponds
- In general, one third of the wet floc can be obtained by drying the harvested floc. Shade drying of biofloc will be the easiest method for drying, however drying the wet floc in hot air oven at 40°C for 48 hours is also recommended
- The cost of producing 1 kg of dry biofloc would be approximately Rs. 25 to 30 based on the source of water used for the biofloc production

b. Photobioreactor

Photobioreactor system is equipped with acrylic tubes arranged in 4 rows and five columns at the length of 2m/tube a total of 20m length tubes (Fig 4 & 5) is fixed per unit which provides the capacity of 100 litres/unit. The total length of bioreactor extends to the length of 1.38m and height of 1.5m from the ground surface (Fig. 3). Technical specifications of the Photobioreactor is given in the Table-1.

Motor Pump

One hp motor pump is connected between the bioreactor and the acrylic pipe unit which provides the flow rate of 40-70LPM. It comes with the specification of IP55 which is resistant against the dust and water over the body of the motor pump. These

Figure 3. Outer view of the Biofloc production Raceways

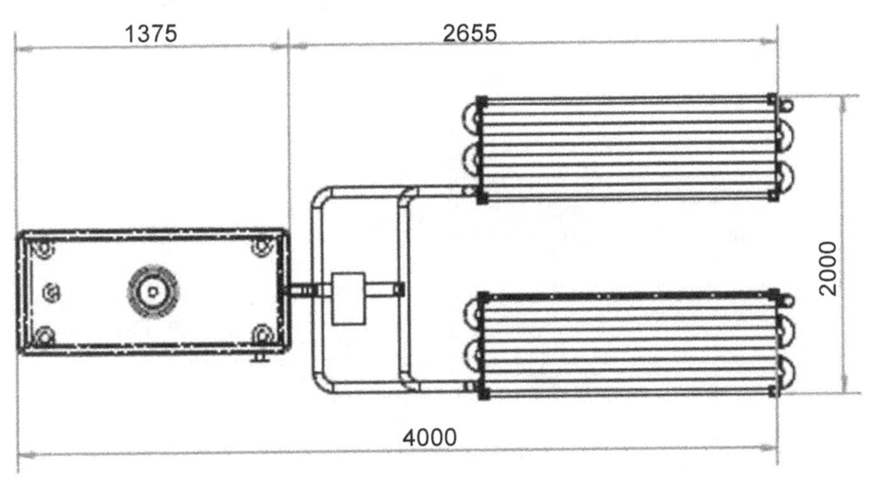

Figure 4. Plan Layout of Photobioreactor (Aerial view)

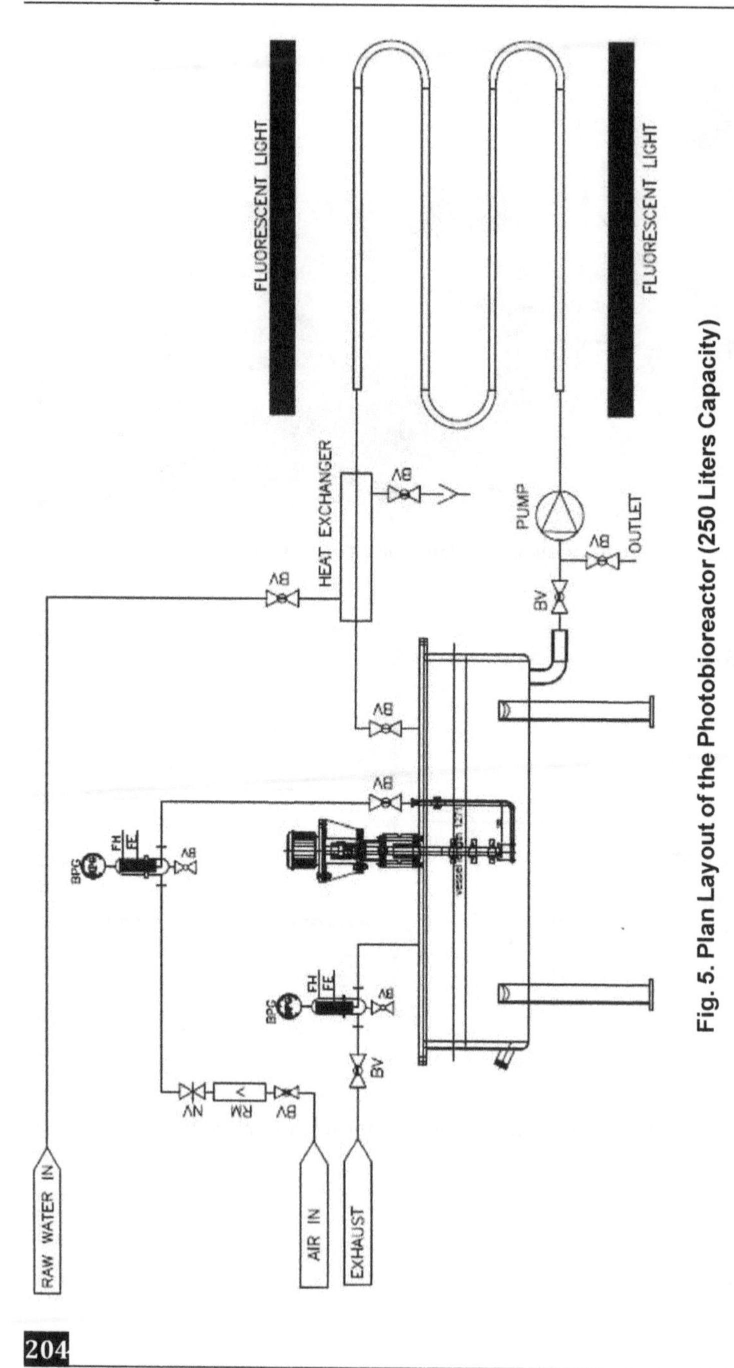

Fig. 5. Plan Layout of the Photobioreactor (250 Liters Capacity)

pumps are capable of operating the bioreactor upto the height of maximum of 20m in the acrylic tubes. Moreover, it runs culture in the anti-gravity flow to maintain the culture to grow at the optimum flow speed range. The proper illumination source with LED lights are fixed to pass through the acrylic tubes for the fast growth of the biofloc culture in the photo-bioreactor (Fig 6).

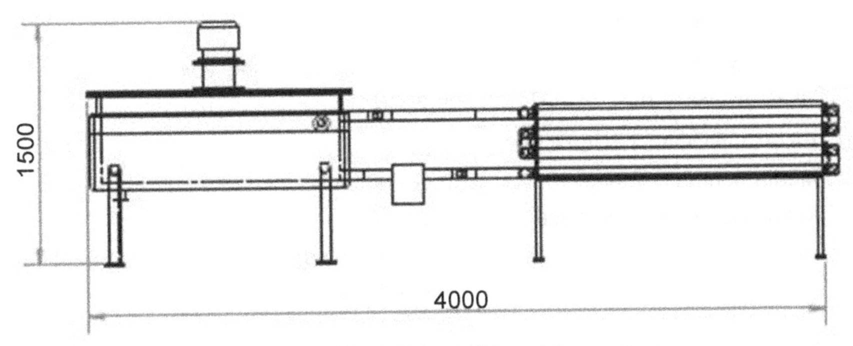

Figure 6. Elevation of Photobioreactor

Agitator

Agitator is installed and connected to the impeller shaft to provide sufficient suspension of the culture and also to avoid the settlement of the denser particles (Fig.7). One hp Tuple motor is installed with an enclosed gear box in the agitator which can run upto the maximum of 250 rpm which is resistant against the dust and water over the body of the agitator.

Solenoid valve

A total of 3 solenoid valve (CO_2, O_2 and water chilling) is installed to the photobioreactor unit. This valve can maintain the pressure from the range of 0.3 to 10 bar. Solenoid valve is connected to the PLC control system so that it can function and maintain the culture automatically. Water cooling solenoid valve is meant for the optimal maintenance of temperature whereas $CO_{2,}$ and O_2 valve is meant for maintaining the optimum pH in the biofloc culture.

Table 1: Technical Specifications of Photobioreactor

250 L FERMENTER-TECHNICAL SPECIFICATION		
VESSEL SPECIFICATION		
1	Total Volume	150 Litrs and Tube Volume 100 Litrs
2	Working volume	200 Litrs
MATERIAL OF CONSTRUCTION		
1	Media contact surface	SS 304
2	Media non-contact surface	SS 304
3	Insulation	Ceramic wool insulation with SS 304 cladding
4	Oring/gasket	Silicon/EPDM
SURFACE FINISH		
1	Inner surface	240 grit mirror finish
2	Outer surface	180 grit satin finish
PORTS		
1	Top plate ports	Addition Port Light Glass Exhaust Pressure gauge Additional port Agitator
2	Shell Top	Sparger, Bypass and view glass assy
3	1/3rd Vessel Port	Port for Dissolved Oxygen, pH & Temperature
4	Vessel Bottom	Outlet Port
5	Jacket	Inlet & Outlet Drain

[Table Contd.

Contd. Table]

AGITATOR		
1	Motor	1 HP
2	Gearbox	30-250 RPM
3	VFD	0.75 KW/1 HP
4	Impeller	Ruston Turbin -3 Nos
5	Mechanical Seal	20 mm Single Mechanical Seal, Dry Running
PIPE RACK		
	Pipe rack construction	Pipe rack consist with following pipe line, air line, exhaust line,clean steam line, raw steam line & drain line.
CONTROL PANEL		
	Control panel	Inbuilt with PID controller for speed. Temperature and its transmitters
TEMPERATURE MEASUREMENT & CONTROL		
1	Temperature Probe	PT100
2	Temperature Range	0-150
3	Output	4-20 ma Accuracy +-0.1 C
4	Temperature controller	Manual control valve to control fermentation. Temperature and sterilization process
SPEED MEASUREMENT & CONTROL		
1	Drive	Variable frequency drive 1 HP AC
2	Input	230V, Single phase, 50 HZ
3	Output	230V, Single Phase, 0-100 HZ
4	Display	Digital indicator for speed

Rotometer

Rotometer functions to adjust the pressure (LPM) of atmospheric air inside the bioreactor and can be managed with the operation of control valve.

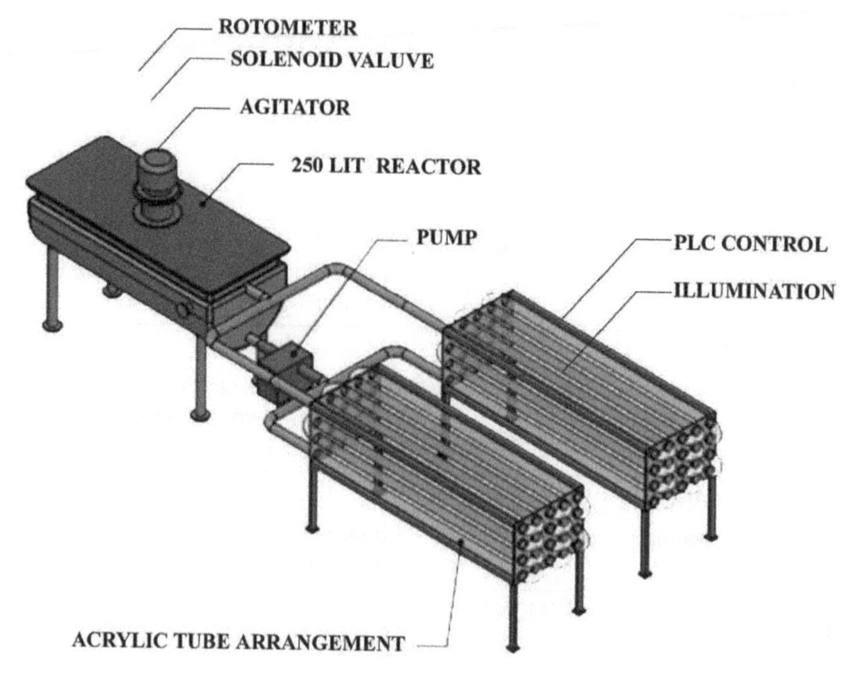

Figure 7. Isometric View of Photobioreactor

Illumination

Photo-bioreactor can utilize the source of both solar and the artificial light source. Supply of light to the photobioreactor can be maintained continuously by integrating artificial solar light devices. A range of 8000 illuminous is provided to the bioreactor to optimize the growth of the culture (Fig. 8A).

PLC Control

Total working function of the bioreactor can be observed in the display and can be controlled automatically. A total of 10 plugins are connected accordingly for various operating functions of the photobioreactor. Data can be archived in the PLC system and can be utilized further when required (Fig. 8B).

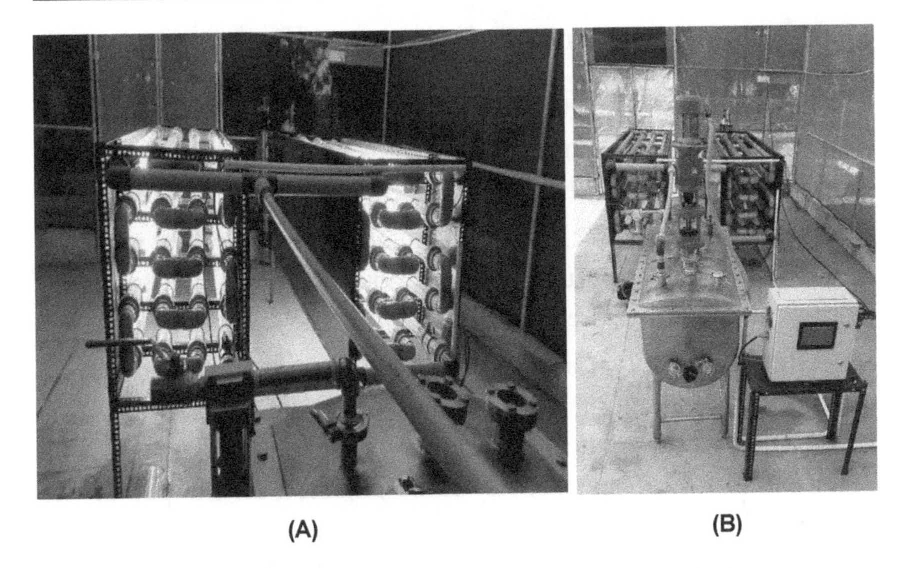

(A) (B)

Figure 8A & 8B. Illuminated Photo Bioreactor & PLC system

Housing of Photobioreactor Unit

The total spacing area of photobioreactor is 6.1×9.1 m. A total of 12 GI pipes with a height of 3.66m and in that 0.6m is fixed inside the ground with concrete mix to make the basement of the roof strong. A total of 5 runners were placed in the arch. Arch's GI pipes have the thickness of ½ inch too. Two doors are fixed in the photobioreactor construction one in the front side and the other in the back side. Side GI pipes are placed at a distance of 3m each and the front door pipes are placed with a distance of 2.4m on the side and 1.22m at the centre. MS metal is fixed on the sides of the photobioreactorto provide the required support.

Protocol for Biofloc Production in Photobioreactor

- Nitrogen sources are mainly nitrate and ammonia obtained from culture water and carbon source used are CO_2, HCO_3^- or organic carbon, like acetate or glucose.
- The air stripping technique is applied to remove total nitrogen and total phosphorus from the aquaculture effluents before inoculating into the photobioreactor.
- After the purification of the pollutants, pure culture of biofloc is inoculated along with growth medium. Gas mixture is supplied to the photobioreactor using sparger and air flow rate and CO_2 flow rate has to be adjusted on the pH of the media.
- The culture is circulated through the tubes by an airlift pump or other suitable low-shear mechanism. The maximum flow rate is limited by the tolerance of the algae to hydrodynamic stress. The flow velocity is usually 0.3–0.5 $m\ s^{-1}$.
- Physical parameters such as pH, ORP, DO and temperature have to be monitored throughout the culture process. A PLC 8.5 software has been used to acquire the data from the transmitters and record the data.
- The biomass measurement can be recorded for every 12 hours and a volumetric pipette can be used to withdraw 100 ml sample from the growing media. Then the average biomass on a daily basis can be calculated to compare the values with the 13L reactor.
- Biofloc biomass can then be separated using vacuum filtration equipment and the concentrated biomass can be collected using Whatman filter paper (number 4) with a diameter of 40mm and the pore size of 20-25μm.

- The biomass collected on the filter paper can be weighed and dried in hotair oven at 70 to 80°C for 10 hours to dry the biomass. The oven dried biomass can be used as an ingredient for the fish/shrimp feed.

Use of photobioreactor and in-pond raceway for the production of biofloc is considerably a new approach in aquaculture. The importance of this approach lies in replacement of conventional ingredients in fish feed as well as for the upkeepment of environment. The designs discussed for the *ex-situ* biofloc production can serve as a model which clearly brings out the advantage of incorporation of biofloc meal in fish/shrimp feeds. The operational experience with the new systems being built with this configuration will allow continuous progress in the knowledge of design criteria and parameters to be used in the *ex-situ* biofloc production.

A comparison of advantages and disadvantages of In-pond raceways and Photobioreactor is presented in the Table 2.

Table 2

Factors	In-Pond Raceways	Photo-bioreactor
Capital cost	Smaller	Larger
Difficulty in terminating nitrification	Greater	Smaller
Operational costs	Smaller	Greater
Operational problems	Smaller	Greater
Process control equipment	Smaller	Greater
Sludge production	Smaller	Greater
Susceptibility to changes in BOD load	Greater	Smaller
Susceptibility to changes in toxic load	Greater	Smaller

References

Burford, M.A., Thompson, P.J., McIntosh, R.P., Bauman, R.H. and Pearson, D.C. 2004. The contribution of flocculated material to shrimp (Litopenaeus vannamei) nutrition in a high-intensity, zero-exchange system. Aquaculture 232: 525–537.

Collazos-Lasso, L.F. and Arias-Castellanos, J.A. 2015. Fundamentos de la tecnología biofloc (BFT). Una alternativa para la piscicultura en Colombia. Revisión Orinoquia, 19: 77–86.

Crab, R., Defoirdt, T., Bossier, P. and Verstraete, W. 2012. Biofloc technology in aquaculture: Beneficial effects and future challenges. Aquaculture, 356: 351–356.

Kumar, S., Anand, P.S.S., De, D., Sundaray, J.K., Raja, R.A., Biswas, G., Ponniah, A.G., Ghoshal, T.K., Deo, AD, Panigrahi, A., and Muralidhar, M. 2014 Effects of carbohydrate supplementation on water quality, microbial dynamics and growth performance of giant tiger prawn (Penaeus monodon). Aquac Int 22: 901–912

Kuhn, D.D., Boardman, G.D., Craig, S.R., Flick, G.J. and Mclean, E. 2008. Use of microbial flocs generated from tilapia effluent as a nutritional supplement for shrimp, Litopenaeus vannamei, in recirculating aquaculture systems. J. World Aquacult. Soc. 39, 72–82.

Kim, S.K., Pang, Z., Seo, H.-C., Cho, Y.R., Samocha, T. and Jang, I.K. 2014. Effect of bioflocs on growth and immune activity of Pacific white shrimp, Litopenaeus vannamei post larvae, Aquac.Res.45, 362-371.

Tacon, A., Cody, J.J., Conquest, L.D., Divakaran, S., Forster, I.P., Decamp, O.E. 2002. Effect of culture system on the nutrition and growth performance of Pacific white shrimp, Litopenaeus vannamei (Boone) fed different diets. Aquacult. Nutr. 8: 121–137.

Wasielesky,W., Atwood, H., Stokes, A. and Browdy, C.L., 2006. Effect of natural production in a zero exchange suspended microbial floc based super-intensive culture system for white shrimp Litopenaeus vannamei. Aquaculture 258: 396–403.

Xu, W.J. and Pan, L.Q. 2013 Enhancement of immune response and antioxidant status of Litopenaeus vannamei juvenile in biofloc-based culture tanks manipulating high C/N ratio of feed input. Aquaculture 412: 117–124.